Henry Barham

Hortus americanus

Containing an Account of the Trees, Shrubs, and Other Vegetable Productions, of

South-America and the West-India Islands...

.

Henry Barham

Hortus americanus
Containing an Account of the Trees, Shrubs, and Other Vegetable Productions, of South-America and the West-India Islands...

ISBN/EAN: 9783337058098

Printed in Europe, USA, Canada, Australia, Japan

Cover: Foto ©berggeist007 / pixelio.de

More available books at **www.hansebooks.com**

HORTUS AMERICANUS:

CONTAINING

AN ACCOUNT OF THE

Trees, Shrubs, and other Vegetable Productions,

O F

𝔖𝔬𝔲𝔱𝔥-𝔄𝔪𝔢𝔯𝔦𝔠𝔞 𝔞𝔫𝔡 𝔱𝔥𝔢 𝔚𝔢𝔰𝔱-𝔍𝔫𝔡𝔦𝔞 𝔍𝔰𝔩𝔞𝔫𝔡𝔰,

AND PARTICULARLY OF THE

ISLAND OF JAMAICA;

Interspersed with many curious and useful OBSERVATIONS,
respecting their USES *in*

MEDICINE, DIET, AND MECHANICS.

BY THE LATE

Dr. *HENRY BARHAM.*

TO WHICH ARE ADDED,

A LINNÆAN INDEX, &c. &c. &c.

KINGSTON, JAMAICA:
PRINTED AND PUBLISHED BY ALEXANDER AIKMAN, PRINTER
TO THE KING'S MOST EXCELLENT MAJESTY, AND
TO THE HONOURABLE HOUSE OF ASSEMBLY.
MDCCXCIV.

T O

THE HONOURABLE WILLIAM BLAKE, ESQUIRE,

SPEAKER,

AND THE OTHER MEMBERS OF THE HONOURABLE

HOUSE OF ASSEMBLY,

THIS ATTEMPT TO RESCUE FROM OBLIVION

THE REMAINS OF AN ANCIENT AND RESPECTABLE

WRITER OF THIS COUNTRY,

IS MOST RESPECTFULLY INSCRIBED,

B Y

THEIR VERY OBEDIENT,

AND DEVOTED SERVANT,

A. AIKMAN.

P R E F A C E.

IT would be doing injuſtice to the memory of the late doctor Barham, to ſuppoſe it neceſſary for the editor to make any formal apology for publiſhing what is univerſally allowed to be the genuine performance of ſo reſpectable a character, and which, from its own intrinſic merit, cannot fail of recommending itſelf to every reader.

Although no doubt can be entertained of the genuineneſs of the work, yet it muſt be owned that there is much appearance of want of exactneſs, and ſuch perfection as might be expected from his hand : This, however, muſt be attributed to the length of time which has elapſed ſince the death of the author, and the numerous hands through which the manuſcript has paſſed.

Sir Hans Sloane, in the Appendix to his ſecond volume of the Natural Hiſtory of Jamaica, ſpeaks in ſtrong terms of doctor Barham, and has made copious extracts from the work, which he ſays he received in manuſcript from the doctor,

tor, and which he expreffes a wifh may foon be publifhed: The editor, however, is not aware that any other part, excepting the quotations made by Sir Hans Sloane, and fome extracts interfperfed through Mr. Long's *Synopfis*, was ever prefented to the public.

The anxiety expreffed by many to fee the entire work of doctor Barham, has induced the editor to aim at rendering it as complete as poffible, by the addition of a Linnæan index, for which he is indebted to a gentleman eminent for his botanical knowledge; an index of difeafes, remedies, &c. has alfo been annexed, from which reference may readily be had to any part of the work, and, it is prefumed, in a manner intelligible to all claffes of readers.

It was the wifh of the editor to accompany the work with fome account of its ingenious and philanthropic author; but his refearches and enquiries have not produced any materials wherewith to gratify curiofity. All he can learn is, that he came to this country early in the prefent century, and married Elizabeth Fofter, the widow of Thomas Fofter, efquire, of St. Elizabeth's, in whofe right he became poffeffed of a confiderable fortune in that parifh; he afterwards purchafed of the family of the Stevenfons, relations of Mrs. Barham, Mefopotomia eftate, in Weftmorland. In the jour-

nals

nals of the affembly, we find him a member of that body in the year 1731; and it appears that he returned to England about the year 1740, and fettled with his family at Staines, near Egham, where he died, leaving his property in this ifland to Jofeph Fofter, the youngeft fon of Mrs. Barham by her former hufband, on condition of his affuming the name and bearing the arms of Barham, in addition to thofe of Fofter. This gentleman was the father of the prefent Jofeph Fofter Barham, efquire, a member of the Britifh parliament, and lately married to lady Caroline Tufton, daughter of the earl of Thanet,

HORTUS

HORTUS AMERICANUS.

ALDER-TREE.

THERE is a fort of alder grows in Jamaica, the virtues whereof are much the fame with the Englilh alder, as cooling, drying, and binding.

ALDER-TREE, or BUTTON-WOOD.

We have a fort of alder, which is commonly called, in Jamaica button-tree. It hath a laurel leaf, and fmall yellow flowers, with alder-like cones. The fruit is drying, binding, and healing.

ALLIGATOR-WOOD. *See* Mulk-Wood.

ALSINES, or CHICK-WEED.

We have three or four forts of thefe. They are cooling, and of the nature of purflanes, &c.

AMBERGRIS.

Many are the opinions about the origin of ambergris, but none hath yet concluded with certainty

A   what

what it is. A certain mafter of a veffel affirmed, he faw a fpermaceti whale opened in North America, in the bowels of which was found a great quantity of ambergris, which made them believe it was the dung of that fifh ; but I am more inclined to believe the whale might fwallow it, meeting with it floating in the fea ; and indeed volumes have fwelled with diverfity of opinions about this reftorative treafure, yet all agree in its juft tranfcendent virtues ; and, let it proceed from what it will, or whatever it be, it is certain that it is a moft powerful antidote againft poifons ; for the Bermudians, thofe of the Bahama iflands, and the Florida Indians, whenever they are poifoned with fifh (which often they are), they fly to ambergris as a powerful antidote, and are cured therewith, and refcued from the moft horrid fymptoms threatening them. People that are acquainted and know the ufe of this fovereign remedy, take it in all weakneffes, and in great evacuations by vomiting and ftools, and in all other too-liberal difcharges of nature and ftrength ; in difpiritments, in fevers, in the hip, or any melancholy or dejectednefs, they happily take ambergris, and that not in a very fmall quantity. I have feen a man, faith Dr. Trapham, take two drachms at a time, without any prejudice, but made him as merry as if he had drank a great quantity of wine. Ambergris, faith he, by its odoriferous parts, unites the fpirits to themfelves, and ftrengthens by fuch addition thofe brifk minifters of life and fenfe, thereby enabling them to caft forth their enemy. The advantage of fuch auxiliaries far exceeds the tranfient inflammatory fpirit, fuch as rum or brandy, which only hurries the fpirits into a brifk motion, the fooner to haften an exit; whereas, our benign and powerful ambergris brings no danger of inflaming the weak fpirits to a confumption, but rather increafeth

creafeth the ftock; not fpurring nature to an over-ha-
zarding ftrain, but fuccours with adventitious and lafting ftrength, conveying apparently by a lafting and
fubtle power, it being only diffolved in wine, broth, or
other warm liquids; the which when received refrefhes
it, and, by its nervous parts, fendeth impreffions into
the whole body, and refrefhes the whole economy of
nature, ejecting all morbific taints, not only egregious
poifons, but all other noxious and offending humours.
It is a fignal remedy for the horrid fpafms, or lofs of
the ufe of the limbs in the dry belly-ache; it alfo ftops
vomiting and loofeneffes, is proper for all inward
bruifes, and a moft univerfal cordial; it refrefhes the
memory, and eafes all pains of the head, being diffolved
in a warm mortar and mixed with ointment of orange-
flowers, anointing the head, temples, and forehead,
therewith; it alfo helpeth barrennefs proceeding from
a cold caufe, and cures fits of the mother inwardly taken: And Dr. Trapham concludes with faying, that
we dare affign ambergris to be the moft univerfal cordial in the world.

AMBER, LIQUID. See Liquid Amber.

ANCHOACA, a yellow Mallow. See Mallows.

ANCHOVY-PEAR.

This is the fruit of a very large tree, growing very
plentifully in the mountains in Jamaica, and other
parts of America. It hath a leaf above a foot long, and
above half a foot broad, very nervous and tough. Its
fruit is about the bignefs of an alligator or crocodile's
egg, and much of the fhape, only a little more acute
at one end, of a brown ruffet colour; and, when pickled,
exactly refembles the mango, if not the fame thing.

ANGELYN-TREE.

These trees grow in most parts of America. Sir
Hans Sloane calls them *melanomma* and *melanoxylon
arbor laurifolia nucifera gemmis nigricantibus Americana*. Piso calls it *andira* and *ibariba,* five *angelyn,*
p. 175. *See his figure.* He makes two forts. Both
bark and fruit are as bitter as aloes; a scruple of the
bark given in a proper vehicle kills worms, but if you
exceed the dose you may do harm.

ANOTTO.

This plant hath many names, as *urucu, roucou, rocour,
orleana* feu *orellana, ouroucou:* Tournefort calls it *mi-
tella Americana maxima tinctoria,* and so doth Plumier: Hernandez and the Indians call it *achiotl,* feu
medicina tingendo apta.

The figure of the plant, with its flower and fruit, is
extraordinarily well designed in Piso.

The leaves are cordated, or in the figure of an heart,
about four inches long and about two broad, coming
out alternately from the stalks and branches, having a
fort of foot-stalk, and a nerve running through the
whole leaf, with transverse or oblique veins on each
side; at the ends of the branches come out, upon a
short foot-stalk, many flowers in clusters, every flower
the bignefs of a small rose, with five leaves of a carnation colour, with a great many yellow stamina, or
thrums, with purple tips; after the flower follows the
fruit, or cod, which is in the shape of the leaf, but not
fo broad, covered with a very rough coat, like the chefnut, which is first green, and, as it ripens, grows of a
dark brown, and then opens of itself. Every cod contains about thirty or forty feeds, about the bignefs and
shape of buck-wheat, having a splendid red colour, and
a little

a little oily; fo that it tinges or paints the fingers of a reddifh colour, not eafily got out with wafhing; and it is what fticks to the outfide of the feed which makes the pafte called anotto; which they get by wafhing it off with water, and after feparate the water and make the pafte up into balls. This the dyers ufe to make a colour they call Aurora. I have known it fold in America for nine fhillings *per* pound, but now of low price, and much out of ufe.

There is a magiftery prepared with the pafte, as followeth : *Take fine flour of caffada, orange-flower water, white fugar, Brafil pepper, and the flowers of* nhambi, *all finely mixed. (See more of the preparation in Pifo, p.* 116.) This magiftery is given to perfons that are poifoned, in waftings and confumptions, hectic fevers, and immoderate fweatings; it ftops bloody fluxes, ftrengthens the ftomach, and provokes urine and the gravel; there is alfo an extract to be made out of the roots, which is of the fame nature as the pafte. Anotto is commonly put in chocolate; and the Spaniards mix it with their fauces, and broths, or foups, which gives them a faffron colour, and a pleafant tafte.

APPLES.

There are feveral forts of wild fruits refembling the fhape of apples, but are in no refpect like the Englifh apples. There is a fort growing amongft the Bahama Iflands, called feven-years apples, which are indifferently pleafant and fweet, and when ripe are black and full of feeds. They will at firft purge them that are not ufed to eat of them, and afterwards bind ftrongly. *See* Cuftard-Apple.

APPLES of love. *See* Tomato-berries or nightfhade.

APPLES caufing madnefs. *See* Solanums or nightfhade.

APPLES, thorny. *See* Thorny apples or nightfhade.

ARAQUIDNA. *See* Pindals.

ARRAGANAS. *See* Myrtles.

ARROW-HEAD.

This grows in great plenty in Jamaica. Sir Hans Sloane faith, he hath feen the fame plant fent from Fort St. George, in the Eaſt-Indies, by the name of *coolette yella*. It grows much like our European arrow-head, and hath its name from its fhape; *viz. fagitta* five *fagittaria*. Tournefort calls it *ranunculus paluſtris folio fag ttato maximo*. It generally grows in ſtanding waters, and is counted a peculiar wound herb, whether inwardly taken or outwardly applied; the root, bruiſed and applied to the feet, helps the crab-yaws in negroes.

ARROW-ROOT.

This root is fo called from its curing and expelling the poifon which Indians put to their arrows when they fhoot at their enemies, which, if they make but a ſlight wound, certainly kills the perfon if the poifon be not expelled; and that this plant doth, by taking the juice inwardly, and applying the bruiſed root as a poultice outwardly: This was difcovered by an Indian, taken after he had wounded an European with one of thefe poifoned arrows, whom they tortured until he promiſed to cure him, which he did effectually with the root of this plant. It hath a ſtalk and leaf exactly like Indian ſhot, only that hath a beautiful fcarlet flower, and this hath a milk-white one. The leaves of it fall in December, and the root is fit to dig in January. Sir Hans Sloane calls it *canna Indica radice alba alexipharmaca*, from its known virtues in expelling poifon. I knew a gentlewoman in Jamaica that was bit or ſtung with a black fpider (which is venomous here) upon one of the

fingers,

fingers, which immediately inflamed and pained her
up to the elbow and fhoulder, and threw her into a
fever, with fymptoms of fits; and all this happened in
lefs than an hour. They fent away for this root, which
they took and bruifed, and having applied it to the part
affected, in half an hour's time fhe found much eafe;
in two hours afterwards they took that away, and applied
a frefh root, which ftill brought more eafe and quiet-
nefs of her fpirits; her fever abated, and in twenty-four
hours fhe was perfectly well. I knew another perfon
cured in the fame manner, that was bit by one of thefe
fpiders, at the neceffary-houfe, upon the buttock: And
about three miles from St. Jago de la Vega, happened
an accident of poifon not defigned, which was done by
an ignorant negro flave, by ftopping a jar of rum with
a weed, which will be defcribed hereafter. The rum
ftood ftopped all night, and fome of the leaves had
fallen into it: In the morning, a negro drank of it,
and gave fome to two or three more of his country;
and in lefs than two hours they were all very fick
with violent vomiting and tremblings. This alarmed
the plantation, and the mafter of it was fent for, let-
ting him know that fome of his negroes were poifoned,
but how they could not tell. He took a furgeon with
him; but before he got there, two or three of them
were dead, and another juft expiring. The furgeon
was at a ftand what to do; but fomebody advifed
Indian arrow-root, which they got immediately, and
bruifed it, being a very juicy root, and preffed out
the juice, and gave it to the negro, who was feemingly
a-dying: The firft glafs revived him, the fecond brought
him to himfelf, fo that he faid he found his heart *boon*,
and defired more of it; upon which he mended, and
in a little time recovered. This is Lopez de Gomara's
counter-poifon, and is one of the ingredients of Her-

nandez'

nandez's grand elixir, or great antidote. I have feen this root frequently given in malignant fevers with great fuccefs, when all other things have failed. When I make up *lapis contrayerva* for my own practice, I always put in a good quantity of it. I have given it decocted, but it is beft in powder, which caufes fweat; the dofe is from a drachm to two. I have obferved, that although this is a very flowery root, yet, if you keep it feven years, no vermin will meddle with it, when all other roots in this country are very fubject to be deftroyed with worms and weevils. It hath no manner of ill tafte or fmell; it works by fweat and urine, and yet is a great cordial; it provokes the terms, and clears lying-in women; it drives out the fmall-pox or meafles; and if it was candied as eringo-root, it would make a pleafant preferve, for it poffefes the like prolific virtues.

ARSMART.

We have two forts of arfmart in America, the fame as grow in England, one without fpots, the other with. It is known, as the great and learned Boyle commends it, as a fpecific to break the ftone and expel the gravel in the reins or bladder, and that by a fimple water diftilled from this plant; but its juice or effence, in my opinion, is much better, fweetened with a little fyrup of marfh-mallows. The root, bruifed and applied to an aching tooth, takes away the pain; the juice or effence, mixed with equal quantities of ox-gall, oil of fpike, and muftard, well mixed, difcuffes all cold fwellings, fcrofulous and fchirrous tumors, and whitlows or felons; the effential oil is good for knotty gouts; or this: *Take the oil of arfmart (made by infufion), lovage, and fhepherd's purf, of each a handful; the head of five fheep and fifteen frogs; boil all together in two or three quarts of oil, until the flefh is*
confumed,

confumed, and then ftrain. This is excellent for knotty or chalky gouts, rubbing it well into the parts.

ASPARAGUS.

The common garden afparagus never grows fo large in Jamaica as they do in England. We have a'fo a fort of wild fea afparagus: It is a more powerful diuretic than garden afparagus, befides having all its virtues.

ATTOO.

I never could find any other name for this plant, and that I had from a negro. I take it to be the fame plant that Sir H. Sloane calls, in his catalogue of Jamaica plants, *radix fruticofa glycirrhizæ fimilis cortice fufco, &c.* and indeed the root to the fight much refembles Eng'ilh liquorice, but of a bitterifh tafte. It hath leaves like the dogwood tree, but is a fmall fhrub, hard'y able to fupport itfelf, and generally joins to another plant, although it doth not climb about it; it hath a fhort pod, which when ripe is very black and full of fweet pulp, like *caffia fiftula.*

The negroes cleanfe their teeth with this root; and they alfo grind it with water like a pafte, and plafter their bodies all over with it in moft feverifh heats, headachs, and cholics; and have fuch an opinion of it, that if they find not a prefent relief by it, they give themfelves over. A certain gentleman recommended it to me as an excellent remedy in the dry belly-ache; and I happening to have a fervant feized with it, to that degree as threw him into convulfion fits, I thought fit to make ufe of it, by decocting the root, and giving him about half a pint at a time, warm, three or four times a-day; which firft eafed him of all his pains, afterwards wrought gently downwards, and, in three or four days,

he

he said he thought himself as well as ever he was in his life, and so continued.

AVENS.

There are two or three sorts of them growing in America. One sort, Pere le Feuville calls *caryophylata folius alatis flore amplo coccineo*. It is an aperitive herb, which the natives make a tea of, to keep their bodies in order. It grows about half a yard high, on the side of the mountains, and hath a scarlet blossom. The same sort I found growing in Jamaica: It is hot and dry, attenuates, cleanses and opens obstructions; is good in bruises and pleurisies, and heals wounds.

AVOCADO-PEAR.

This tree and fruit are well known in America; in the kingdom of Peru they are called *pattas*.

The fruit is of a pear fashion, as big as the English pound pears, and green when ripe; but I have seen a sort very round, with red streaks like a pear-main. When they have been gathered some days, they grow soft, and are fit to eat with pepper and salt; some mix them with lemon-juice and sugar, others will boil them and eat with salt beef. They are very nourishing, and are thought to be great provocatives; therefore the Spaniards do not care their wives should eat much of them. This fruit is ripe in June, and so continues till October. They have a large stone in the middle, wrapped up in a fine thin skin, of the shape of a heart; and when that skin is taken off, it is very rough, and in wrinkled or little hard protuberances, of a reddish colour; when cut through, it is very white; but the air soon turns it reddish. If you take one of these pear-stones, and write upon a white wall, the letters will turn as red as blood, and never go out until the wall is white-washed.

washed again, and then with difficulty; also, if you take a piece of white cloth and put round them, and with a pin prick out any letter or figure on the cloth, the figure will be of a yellow colour, not to be easily washed out.

BALSAMS and GUMS.

See, in the order of the alphabet, Liquid amber, Ambergris, Gum animi, Gum cancamum, Gum caranna, Bdellium, Balsam capaiba, Copal, Elemi, Balsam nervinum, Balsam Peru, Hog-gum, Balsam Tolu, Tacamahac, Dragon's blood, Gamboge.

BALSAM CAPAIBA.

This balsam is called by several names; *viz. capivy, copahu, copau, copalyva, capaif,* and *campaif;* by the native Indians *colocai;* by the Brasilians *copaiba;* by the Portuguese *gamelo* or *gamemolo.* Many Americans, particularly the Mexicans, do call all resins and sweet-smelling gums or balsams by the name of *copal,* although there is a gum that is more particularly so called, which I shall describe hereafter.

The wood of this tree is red; the leaves are four or five inches long, and oval, with small stems and hard nerves on the back; the flowers are pentapetalous, or five-leaved, standing round the fruit or pods, which are roundish, with a thin black shell, when ripe or dry easily broken, containing a small yellow smooth pulp, smelling like pease, which the Brasilians suck the juice of, and spit out the skin; the monkies also are very greedy of them. They begin to ripen in April, and are full ripe in June.

To get the liquor or balsam, you must bore the tree to the pith at the full moon, which will run in such quantity that in three hours it will run sometimes twelve

twelve pounds.   But if it fhould happen that little or
none fhould run out of the wound, then they ftop it
up immediately with a plug, luteing it with wax, that
nothing may flow out; and, after fourteen days, it will
compenfate the delay with intereft   This tree is not
fo common in the Prefect of Parnambuca as in the Ifle
of Maragnan, and about Surinam and the Capes,
where it is plenty, and from thence we have it in great
quantities.

I have feen another way of getting the balfam, which
is by cutting the tree halfway through, the upper part
cut floping, the lower part ftrait in; and when you
have cut enough, dig the lower part like a bafon, fo
the balfam will drop very faft, and, as that fills, lade
it out into proper veffels; by this method, I have known
barrels of it filled in a little time; and it is fo plentiful in fome places, that they burn it in their lamps in
the room of oil.   It doth not fmell fo pleafant at firft
as it doth afterwards, and is clearer and yellower when
old, and thicker, &c.   The natives found out fome
of its virtues by the wild boars or hogs running to the
tree when wounded, ftriking their tufks againft the
trunk, and the balfam, flowing out into their wounds,
perfectly healed them.

It is certain that the balfam *capivy* is a moft admirable medicine, either internally taken or externally applied: It is a fpecific in the gonorrhœa, after due purging, and the whites in women; it alfo cures coughs
and confumptions of the lungs; it is hot and bitterifh
to the tafte, and of an aromatic fmell, very clear and
tranfparent if right good, and not much inferior to balm
of Gilead; and is the fame which they make fuch a
great noife about, under the name of balfam Chili.   It
comforts and warms a cold ftomach and bowels, and
is excellent in cholics or belly-aches; by its fubtle penetrating

netrating parts it enters into the whole mafs of blood, depurating it, provoking fweat, and forcing urine, powerfully opening all obftructions; it is a great vulnerary, curing wounds even of the nerves. You may mix it with fugar, oil of rofes, and plantain water, and ufe it as an injection, to heal ulcers in men or women; it is alfo good in a clyfter for the belly-ache.

BALSAM-HERB.

This herb is fo called in Jamaica, and few or none know it by any other name, although it is a fort of *antirrhinum*. This in Jamaica fmells, when rubbed in the hand, almoft like *melilot*, or fome pleafant balfam; and therefore they call it balfam-weed or herb, and make a balfam of it. The juice or diftilled water is good for fore eyes.

BALSAM NERVINUM.

This balfam is made after the manner of oil of bayes, by boiling a fmall red cluftered fruit or berries of a large tree, whofe leaves are very large and broad and green; they grow in great plenty in St. Domingo and other ifles. This balfam is in colour like Tolu, but of a lefs agreeable fmell.

BALSAM PERU.

The tree that this balfam comes from is the *cabureiba* of Pifo, of which there are two forts, very high and large. The one fort hath a reddifh bark, and fmells like cedar; the other fort hath a fmall leaf like myrtle, the bark of an afh colour, very thick, but the outward coat or fkin is very thin and reddifh, under which lies the yellow liquor or balfam, which, when old, fmells much more fragrant, growing thicker and redder when there is a frefh fpring in the tree, which is

about

about February or March, and at the full of the moon. They get this fragrant balfam out of the wounded bark, and receive it into calabaſhes. It is certain this balfam is excellent for wounds of the nerves, and reſolves cold tumors; inwardly taken, it ſtrengthens the ſtomach, reins, and back, and drives out malignant humours by perſpiration. Some get this balfam by boiling its bark, branches, and leaves in water, ſkimming off the top; but this is a very black fort: The beſt fort is of a blackiſh-red colour, and is always liquid, of a ſweet agreeable taſte, ſmelling like ſtorax or cition; or rather vanilloes when well cured. It is uſed as a great pectoral, particularly in aſthmas.

BALSAM TOLU.

This balfam hath its name from a little village called Tolu, ſituated near Golden-Iſland, or the Stockadoes, in Darien, near Nombre de Dios, near where the Scotch took poſſeſſion in King William's reign. It is fold in calabaſhes; becauſe, as it grows old, it grows reſinous and brittle. It is of a grateful fragrant ſmell, a great pectoral, particularly in phthiſicks, catarrhs, and defluxions, made into a ſyrup, which you may make very pleaſant and fine, in the following manner: *Take four ounces of balſam, putting it into a flaſk, filling it with water about two-thirds full; then put it cold in a veſſel of water, and let it gently boil for twenty-four hours; then pour off the clear, to which add double its weight of double-refined ſugar, and make a balſamic ſyrup.* What is not diſſolved, may ſerve again to make more ſyrup.

BALSAM-TREE.

This tree is fo called becauſe fo much balfam comes from it, even from the bark, leaves, and fruit. Sir
H. Sloane

H. Sloane tribes it amongſt his *terebinthi*, or turpentine trees; but it is in no reſpect like any of the fir
kind, it is certain. It hath very thick, round, and
brittle leaves, and, when broke, comes out a milky
juice, which immediately turns yellow, and ſticks to
the fingers like bird-lime; the fruit is the bigneſs of a
genetin, or Indian wild fig, and full of gum. If you
cut the bark of the tree, immediately comes out a yellow gum, but without ſcent. I queſtion not but the
gum would be of great uſe, if experienced; for we
know not as yet the virtues of it, nor ever could meet
with any that could give me any medicinal uſe of it;
if the Indians know, they keep the uſe of it to themſelves. They grow in great plenty in Jamaica; and
are ſo plentiful in moſt parts of America, that in ſome
places they mix this gum-juice with tallow, and paint
their canoes and boats with it, to make them glide
through the water, and preſerve them from worms.

Banana-Tree.

This is very common, and its fruit ſo well known
that it needs no deſcription. The Spaniards have a
conceit, that if you cut this or the plantain athwart or
aroſſways, there appears a croſs in the middle of the
fruit, and therefore they will not cut any, but break
them. The Franciſcans dedicate this fruit to the
muſes, and therefore call it *muſa*. The Portugueſe call
them *ficus derta*, others *ficus martabana*; in Guinea,
bananas. Lodovicus Romanus, and Brocard, who
wrote a deſcription of the Holy Land, call them Adam's
apples, ſuppoſing it to be the fruit that Eve took and
gave to Adam, which is erroneous; but it is very probable, that their leaves might be the fig leaves they
ſowed together to hide their nakedneſs; nay, one leaf
alone was or is ſufficient to do that, being very broad
and

and long; I know none like it. They are a wholefome
fruit, and make a pleafant drink, exceeding Englifh
cyder; baked, they eat like an apple, and fo they do
in a dumplin; dried in the fun, they eat like a delicate fig. The juice of the leaves is good againft a
burn; the fruit comforts the heart, and cools and refrefhes the fpirits; made into a marmalade, or comfit,
it is good for coughs and hoarfenefs, lenifies the fharpnefs of humours defluding upon the lungs, and allays
the heat of urine.

See Plantains.

Barbadoes Flower Fence.

This, I fuppofe, is fo called from their fencing in
their plantations with this fhrub, which is full of fhort
ftrong prickles; but they are commonly called in Jamaica *doodledoes;* they grow in all or moft parts of
America. The flowers are elegantly mixed with red-
yellow, and therefore called, by fome, Spanifh carnation, or wild fenna. Sir Hans Sloane tribes it amongft
the baftard fenna's, for this comes the neareft of any in
America, and, when dried and old, it is very difficult
to diftinguifh one from the other; and as for virtues,
I have often experienced it to have the fame with that
of Alexandria; befides which, a decoction of the leaves
or flowers has a wonderful power to move or force
the *menftrua* in women. The flowers make a delicate red purging fyrup, and the root dyes a fcarlet colour. The whole plant is full of fhort fharp prickles,
branching and fpreading very large, with beautiful
flowers, red mixed with yellow, on which are a great
number of thrums like faffron; the leaves, when green,
are of the fhape of indigo; the pod is in fhape of the
Englifh broom pods, or like the fenna of Alexandria;
when ripe and dry it is black, containing five or fix flat
feeds,

feeds, cordated, and of a dark-greenish colour. This shrub is fulleft of flowers in the months of November and December, and the feed is ripe in January.

Basil.

We have in Jamaica two or three forts of bafil; but that which grows fpontaneoufly, and moft common, is that fort which Sir Hans Sloane calls *ocymum rubrum medium*. There is another fort in South America, mentioned by Monfieur Frezier, called *alva haquilla;* a fhrub, faith he, which has the fcent of our fweet bafil, and contains a balm of great ufe for fores; whereof we faw a wonderful effect at Yrequin, in an Indian, whofe neck was deeply ulcerated. I alfo had the experience of it on myfelf. The flower of it is long, growing up like an ear of corn, of a whitifh colour inclining to a violet, and is tribed amongft the *legumina*. Bafils are fpoken againft by Diafcorides, Galen, and Chryfippus; but Pliny commends them much, and faith they are good againft the fting of fcorpions and other venomous ferpents, and are accounted a very great cordial, and good againft pains of the head, &c.

Bastard Cedar,

As it is here fo called; for what reafon I know not, being in no refpect like cedar. Its leaves are in the fhape of Englifh hazel; its fruit like the mulberry, firft green, and when ripe black and hard, which fheep and cattle delight to eat, and will make them fat. I take this tree to be of the mulberry kind, more than of the cedar; the flowers are like the line or lindal tree, yellowifh, and very odoriferous, fmelling like our May or hawthorn flowers.

B   Bastard

Bastard Mamee, *or* Santa Maria.

These are very tall trees, and very straight, growing to fifty or sixty, some to eighty feet high; they are very tough, and therefore made use of for masts of ships, being preferable to any fir trees. I had once a green balsam presented to me, brought from the Spaniards, of a very fine green, clear, and pleasant smell; which they said was the finest balsam in the world for green wounds, but could not tell me from what tree it came. Some time after, a negro brought me of the same sort of balsam, both in colour and smell, which he got from one of these trees, and I found it to be an excellent balsam; for, melt it and pour it into a green or fresh incised wound, and it would heal up in once or twice dressing. This balsam the Spaniards, while it is new and fresh, put into the hollow joints of trumpet-wood, calling it *the admirable green balsam*, but conceal its name, and the tree it comes from; yet it is for some extraordinary use that they call this tree *Santa Maria*, which makes me think it is for its balsam.

Bdellium

Is said to flow from the trunk of a tree full of prickles, called *bdellia*. Its leaves are like the oak; the fruit resembles a fig, and is of a pretty good relish; the gum of a bitterish taste, and turns yellow upon the tongue; the best comes in oval drops, is fragrant, reddish, and transparent. It is used both externally and internally, being aperitive, sudorific, digestive, and discussive; it hastens births, provokes terms, and resists poisons. They sell *gum alouchi* for *bdellium*, which is a cheat.

Beans *and* Pease.

The beans and pease of Jamaica are most of them convolvulous

convolvulous plants: The beft fort is the broad bean with blue fpecks. Sir Hans Sloane makes about twenty-one forts growing in Jamaica, including the *bonavift*, white and red fort, the fmall red fort, and the great Angola red peafe, the clay-colour, and the *calavances*, which are all fweet and pleafant, and may be had green all the year round. The horfe-bean and cocoon are venomous, and not to be eaten.

BEAN-TREE.

This beautiful tree grows in plenty in moft parts of America. In the ifland of Jamaica, they make fences of them, being very prickly. About Chriftmas, thefe trees are to be feen all full of large red flowers, without any green leaves, being very beautiful and pleafant to the fight. After the flowers are fallen, the green leaves fhoot out, and the fruit begins to appear, which is a pod about fix or feven inches long, containing about eight or nine beautiful red beans, in the fhape of kidney-beans. The trees are generally very large and fpreading, armed full with black crooked thorns, like cock-fpurs; the leaves are like thofe of the phyfic-nut. The virtues of this plant have not yet been difcovered, though I know by experience that the flowers make an excellent eye-water. Bontius faith, that the fruit is a great diuretic, and purgeth ftrongly water, and therefore proper in dropfies; he faith they expel wind, and cure the cholic.

BELLY-ACHE WEED.

This plant is fo called from curing the belly-ache or cholic with coftivenefs, which was firft made known in Jamaica by Papaw negroes, and therefore commonly called Papaw weed; by this name I knew it. Its leaves and fruit are like the wild cucumber, but

B 2   much

much lefs. It works very ftrongly upwards and downwards, and therefore ought to be given to ftrong perfons, and in the beginning of the belly-ache: It is alfo good in dropfies; while the bowels are ftrong, it may be given in clyfters for the fame intentions.

BIGNONIA.

There are many forts of thefe plants growing in America, having their names from Abbe Bignon. They are more for beauty and fine arbours, than of any medicinal ufe.

BIND-WEEDS.

There is in Jamaica a vaft number of bind-weeds, of the convolvulous kind, with bell flowers. Thofe that are known to be of phyfical ufe, will be mentioned as they come in courfe.

BIRCH-TREE.

It is very common in Jamaica, although I do not take it to be the fame with what grows in England; but it having the very fame fort of bark, makes the Englifh here call them birch-trees. They are much larger here than any I ever faw in England; befides, of thefe, after the bark is off, the wood is very white, light, and brittle; none of the twigs are fo tough as to make rods or brooms of; and the gum that flows from the tree is very odoriferous, white like maftic, and hath an aromatic abforbent tafte. I have often given and advifed this gum to be taken in the *lues venerea* with good fuccefs, after due purging. It is fo well known, that it needs no particular defcription.

BISNAGUS, *or* VISNAGA.

Thefe are well known in New Spain, where they
make

make tooth-picks from them. It is a fort of fennel or chervil; and it is the foot-ftalk of the flower and feed they make ufe of after dinner to pick their teeth.

BITTER-WOOD

Is fo called from its exceffive bitternefs: I think it exceeds wormwood, gall, and aloes. I have feen a handful of the fhavings but juft dipped in water, as quick as thought taken out again, and the water left fo bitter that nothing could exceed it. A trough was made of it to give water to hogs, and, to their owner's furprife, although the hogs were ever fo dry, they would not touch the water. This property of the tree hath not been known very long in Jamaica; and it was difcovered by an accident: It being a very free fort of wood to fplit, light, and white, the coopers had made cafks of it, unknowing its bitternefs, to put fugar in, which was fent to England. Soon after, the owner had advice that his fugar was fo bitter it could not be fold: The gentleman thought it was a trick, or a banter; but, upon a ftrict enquiry, found the occafion of it. Of late, bedfteads and preffes are made of it, to prevent bugs, cockroaches, or worms breeding, as they do in other woods, for none of thefe vermin will come near the wood; neither do the workmen care for working it, it bittering their mouths and throats. It kills worms in the body, helps the cholic or belly-ache, and creates an appetite. The wood of this tree, at the firft cutting, is very white, but turns yellow afterwards. Its bark is like the lance-wood, and its leaves like the Englifh afh.

BLACK MASTICK

Bears a round fruit, as big as a wild fig, and black when ripe like a bully; and therefore is called by fome baftard bully.

BLOOD.

Blood-Flower.

It is fo called from its ftopping bleeding when all other remedies have failed; and is fo well known in Jamaica that it needeth no particular defcription. I knew a gentleman that had fuch a flux of blood, by the piles or *hemorrhoids*, that there was no ftopping it, he himfelf, and all his friends, defpairing of his life. At laft, he was advifed to this flower, which was immediately got (for they grow almoft every where) and bruifed, and preffed out the juice, and was given with a fyringe; by which he was perfectly cured. I had a patient that had a virulent gonorrhea, and after I had carried off the virulence, and began to ufe balfamics and reftringents, I found it would not ftop, and all the medicines I could think of were to no purpofe for above twelve months. At laft he took a decoction of the flowers, leaves, and ftalk, of this plant, twice a-day, for five or fix days, and it made him perfectly firm; and fome years after he told me, that he never had the leaft fymptom of a gleet or any other illnefs attend him in thofe parts. Lately, an ancient gentleman confulted me, who had a gleet upon him many years, which he apprehended was pure weaknefs of the veffels, for he was very well in all other refpects: I advifed him to make a tea of the dried flowers, and drink of it in the room of other tea, and at the fame hours, for a month; in which time, he told me, it made him perfectly well, and faid it was worth its weight in gold, and believed, if a man could make it known in Europe, he would get an eftate by it. I have known many old gleets cured by it; and I queftion not but it may be as ufeful to women, for the *fluor albus*, and other exceffive difcharges.

Boxthorn.

Boxthorn

Hath a white wood, hard and folid like box. The leaves, with twigs, are fet oppofite to one another, which are almoft round, juicy, having two reddifh long fharp prickles rifing by the foot of the leaf. It bears a large purple flower, and a round green fruit of the bignefs of a goofeberry. I have feen fometimes leaves growing out of the fruit. It is of a reftringent quali-ty, and ftops all defluxions of the eyes or ulcers, and heals them.

Brasilletto.

The true Brafil is called Pernambuca, being the place from whence they come in Brafil; the Brafilians calling it *ibirapitanga.* It is a thick large tree, with a reddifh and thorny bark; the leaves fmall and blunt, of a fine fhining green; its flowers little, fweet, and of a beautiful red; the pods flat and prickly, in which are two flat feeds, like the gourd feed. This wood is ufed among the dyers, and the ftationers make red ink of it; viz. *Take rafpings of the wood, infufe them in vinegar or fome ftrong lixivium, and, with gum arabic and allum, put them in a glazed pot, and gently infufe them for fome hours.* Some dye the roots of *althea* with it, to clean the teeth withall. I have met with two forts growing in Jamaica; one every way as red as brafil. It hath a red gum, with a reftringent tafte; its wood is very tough and ftrong; the wheelwrights in Jamaica fay, they make the beft fpokes for wheels. A decoc-tion of the wood ftrengthens the ftomach, abates fe-verifh heats, and takes away inflammations and de-fluxions in the eyes.

Bread-Nut Tree.

Why this is fo called I cannot tell, unlefs it be upon

B 4 the

the account of the wild hogs feeding upon its fruit, which makes them very fat. The leaves are good for horſes. The medicinal qualities are not yet known.

BRIER-ROSE OF AMERICA.

It is a drying reſtringent plant. Its fruit is good againſt ſpitting of blood.

BRIONY.

There are ſeveral ſorts of brionies growing in Jamaica; but the fruit of theſe brionies ſeems to be the ſame with thoſe in England; yet their leaves differ very much: And as they have different names here among the common people, they will be mentioned by thoſe names, as they come.

BROOK-LIME

Differs but very little from that of England in ſhape and virtue.

See Pimpernell.

BROOM-WEED.

This plant is ſo called by the negroes in Jamaica, for no other reaſon, that I know of, than becauſe they make a broom with it, being very tough and ready at hand, growing almoſt every where in Jamaica, even in the pooreſt red land; but it hath no reſemblance to the Engliſh broom, being of the mallow kind, having the ſame ſeed, but a yellow flower, which opens every day exactly at eleven o'clock in the forenoon; ſo that, in the country, I have aſked a planter what it was o'clock, when I thought it was growng near noon, and he would go out and look upon this plant, and tell me. The only medicinal uſe I ſaw of it was, the negro women, when their children were ſcabby or mangy, would

would make a bath of this herb, which would cleanse them, and make them thrive.

BUCK-WHEAT.

We have a sort of climbing or woodbind buck-wheat. This American buck-wheat hath round, red, succulent stalks, by which it winds and turns itself round any tree, rising about seven or eight feet high; towards the top, it puts out leaves alternatively, which are green, thick, juicy, and smooth, in the shape of an heart, about an inch and half long; and towards the top come out flowers, very numerous, in oblong spikes, looking like parsnip seed: In the protuberant part of the flowers lie the seeds. The grains of this plant are hot and dry, and of thin and subtle parts: They are good against hysterics, and are esteemed great provocatives.

BULLY-TREE.

This is so called by the Jamaicans, for its fruit when ripe is as black as a bully or damson, but in shape of a Lucca olive; pigeons feed much upon them, and they make them very fat: Its timber is very strong and lasting. There is another sort, called bastard bully. I remember, after the great fire at Port-Royal in Jamaica, in 1703, jesuits bark was so scarce that we gave four pounds for a pound of it, and some practitioners could not get any for love or money; upon which, they made use of the bark of this tree, for intermitting fevers, with good success, but were forced to give twice or thrice the quantity: Since that, they have have found out a bark that every way answers the ends of the jesuits bark, which I shall mention hereafter.

CACAO,

Cacao.

This beautiful plant and profitable tree grew once, in such plenty in Jamaica, that they valued themselves upon it, and thought they were or should be the richest people in the world; but they soon saw themselves deceived, for a blast at once came upon the trees and destroyed them all, and few or none could ever be got to grow there since; what do grow are generally in plantain-walks, or among shady trees, and in bottoms or vallies sheltered from the north winds. This tree grows in bigness and much resembling the heart cherry tree, the boughs and branches beautifully extending themselves on every side, their leaves being much of the same shape; the flower is very beautiful, and almost of a saffron colour; the fruit proceeds from the body (as the calabash) and shall be full almost all the way from the bottom up to the branches, which are also full of fruit, which is first green, and, as it increaseth its bigness, changes its shape and colour, until they are thoroughly ripe. I have seen two sorts; one very large, as big and almost in shape of a cucumber, but pointed at the end, and of a most delicate yellow or lemon colour, with a little red blush of one side when ripe; another sort not so big, of a fine blueish red, almost purple, with reddish or pink colour veins, especially on that side next the sun; they have on the outside ridges and furrows, with smooth bunches or knobs, as cucumbers have. They are ripe and fit to gather in January and in May, having two crops or bearings in a year. The external husk or rind, which is pretty thick, being broke or cut, there appear the kernels adhering to one another by soft filaments, and inclosed in a white pulpy substance, soft and sweet, which some suck when they take them out of their

shells,

shells, which contain ten, twenty, and sometimes thirty, nuts, almost like almonds.   There is much difference in their largeness and goodness; those at Carpenter's river are the largest, those brought from the Coast of Caraccas next, the smallest are those of Martinico. They are cured in the sun upon cloths or blankets. That which we make our chocolate of is the inside of the nut, encompassed with a thin shell or case, which when taken off, the dry and hard substance looks of the colour of a kidney-bean, with crannies or crevices between them.   They are very apt to mould and decay, if they are not well cured; and, if right good, they are plump, smooth, and oily, and of a bitterish taste when raw.   The oil of this nut is the hottest of any thing known, and is said to recover cold, weak, and paralytic limbs, and to smooth the skin.   This nut is very nourishing, as is daily experienced in the West-Indies, where many creoles live in a manner wholly upon chocolate.   The way of making it is so well known, that I need not describe it.

CALABASH.

I suppose the Spaniards gave the name to this tree, its fruit being as big as a man's head (which they call *calabosh*), but rounder; it is so well known in most parts of America, that it needs no description.   I have seen such difference of the fruit of these trees as to contain from an ounce to a gallon.   When they are green, they are full of white juice, pulp, and seeds, which the cattle eat of in very dry times; but which is said to give their flesh an odd disagreeable taste, and also their milk; but I believe that taste is from a weed called guinea-hen weed, and not from the calabash. It is said that the pulp, if eaten, will make a cow cast her calf, or a mare her colt.   It is certainly known (if

not

not too well known) to be a great forcer of the *men-ſtrua*, and of the birth and after-birth; therefore ought to be very cautiouſly given or taken. I once made a ſpirit from this fruit, which was ſo nauſeous as not to be taken alone. This is a uſeful tree for Indians and negroes to make neceſſary furniture for their houſes, as diſhes, cups, and ſpoons, of ſeveral ſhapes, bigneſs, and faſhion; I have ſeen them made, and finely wrought and carved.

CALAVANCES

Are ſmall peaſe, tribed among the *phaſeoli*. The flower is all white, whereas moſt of the other ſorts of peaſe are blue; the pods are five or ſix inches long, containing a ſmall white pea, reſembling the kidney; they are planted any time when rain or ſeaſons come, and in ſix weeks time are fit to eat green. They are very good and ſweet, green or dry, and eaſy of digeſtion; and therefore proper for a hot climate.

CALTROPPE.

There is a plant in Jamaica which Sir H. Sloane hath given a very exact figure of, in his Hiſtory of Jamaica Plants, which he calls *tribulus terreſtris major flore maximo odorato*.

The greater land caltroppe, with a large ſweet flower, hath a deep root, from which ſpring a great many long trailing branches, ſpreading every way on the ground, a foot and a half long, and are round and juicy, brittle and thick; it hath leaves in pairs; the flowers are of an orange or yellowiſh colour, with five leaves, ſmelling ſweet; then follows a ſmall prickly head, with a proceſs like the crane's bill ſeeds, &c. They are cooling and aſtringent.

CAMPIONS.

CAMPIONS.

The fpecific quality of this plant is againft bloody fluxes, being of a drying and binding quality.

CANES,

The chief of which is that they make fugar from, and therefore called *arundo faccharifera;* it is fo well known to the inhabitants of America, that it needs no defcription; and as for the way and manner of making fugar and of refining it, it would be thought prefumption in me to direct: I fhall only fay, that they are fqueezed or preffed in a mill, between three rollers cafed with iron, and the juice boiled up to fugar. I have obferved, although the juice is very fweet, that a gallon of it will make but one pound of good fugar, and as much molaffes, the reft being water, fcum, and dregs; out of which they alfo make rum, but molaffes makes the beft fpirit: It is alfo obferved, that one hundred weight of fugar makes but about thirty-three pounds of fingle-refined, and about fourteen pounds of double-refined.

Sugar is the effential falt of the plant, which is good for the breaft and lungs to fmooth their roughnefs, therefore good for hoarfenefs and attenuating phlegm; for although fugar feems fweet to the palate, yet there is a great acidity in it; for I can draw from it a fpirit as corroding almoft as *aqua fortis,* and therefore fugar decays the teeth, and makes the gums foft and fcorbutic; if too much ufed; neither is it good for thofe troubled with vapours, hyfterics, or hippo's.

There are two other forts of canes, that grow wild, the one hollow and the other not, but full of pith like the elder: When they fpring up out of the ground, they are boiled, and make one of the beft of pickles, **and**

and will keep with good management two or three years: I think it exceeds the mango.

CAPSICUM PEPPERS.

Thefe only differ from one another, in their fruit, in fhape and colour; fome being, when ripe, red, white, and yellow, and are as follows; *viz.*

1. The common red long fort.
2. The great upright.
3. The leffer ditto.
4. The fmalleft, called bird-pepper.
5. The greateft upright fort.
6. The leffer ditto.
7. The pendulous fort, called bell-pepper.
8. The long olive-fafhion pendulous.
9. The upright ditto.
10. The great long upright.
11. The great crooked or horned fort.
12. The leffer ditto.
13. The forked or double-pointed.
14. The fmall round.
15. The greater round upright fort. ⎫ Thefe are
16. The round cherry-fafhion. ⎪ called goat-
17. The broad crumpled cod. ⎬ peppers, for
18. The fhort round yellow-coloured. ⎪ they fmell
19. The long ditto. ⎪ rank like a
20. The hairy-ftalked fort. ⎭ am-goat.

Thefe are all much of the fame nature. The large hollow fort, called bell-pepper, pickled while green, is an excellent relifhing pickle or fauce for meat; the other fmall red peppers, when ripe, taken and dried in the fun, and then ground with falt and pepper, clofe ftopped in a bottle, are an excellent relifher to fauce for fhould fh, and commonly called kyan butter. All thefe forts of pepper are much more of a burning
heat

heat than white or black pepper. Some punifh their flaves by putting the juice of thefe peppers into their eyes, which is an unfpeakable pain for a little while ; and yet, it is faid that fome Indians will put it into their eyes before they go to ftrike fifh, to m ke them fee clearer.

Thefe peppers ftop vomiting, create an appetite, and ftrengthen the ftomach, if rightly prepared ; fome I have known to fwallow a certain number of them whole, as fome do cudebs, for the pain in the ftomach and cholic ; they powerfully provoke the terms, facilitate birth and after-birth, and are good againſt gravel, or tartarous flimy matter that breeds the ftone in the kidnies or bladder. But I would not advife any perfon that labours under venereal fymptoms, or thofe who are hectical, to meddle with them. When infufed or digefted in fpirits of wine, it takes off much of their violent heating and inflaming quality, and they are then great provokers of urine, curing dropfies. Infufed in oil, they take away the numb palfy, or lofs of the ufe of the limbs ; and, mixed with goofe greafe, refolve impofthumes that come from cold, &c.

Near St. Michael de Sapa, in the Vale of Arica, they cultivate the agi, that is Guinea pepper ; where there are feveral farms which have no other product but this pepper. The Spaniards of Peru are fo generally addicted to that fort of fpice, that they can drefs no meat without it, though fo very hot and biting, that their is no enduring of it, unlefs well ufed to it.

CARAPULLO

Is an herb which grows like a tuft of grafs, and yields an ear, the decoction of which makes fuch as drink of it delirious for fome days, like the Eaft India bangart. The American Indians make ufe of it to

difcover

difcover the natural difpofition of their children: At the times when it has its operation, they place by them the tools of all fuch trades as they may follow, as by a maiden a fpindle, wool, fciffars, cloth, kitchen furniture, &c. and by a youth accoutrements for a horfe, awls, hammers, &c. and that tool they take moft fancy to in their delirium is a certain indication of the trade they are fitteft for.

Cardamon.

We have a plant in Jamaica which grows like the wild ginger, but Sir H. Sloane calls it *cardamomum minus pfeudo-afphodelifoliis;* its leaf is more like *orchis* than *afphodel.* This herb is pectoral, purges phlegm, and expels windy humours, &c.

Cashew.

This tree and fruit are fo well known in America, efpecially in Brazil and in Jamaica, that they need no particular defcription. The ftone of this apple appears before the fruit itfelf, growing at the end in the fhape of a kidney, as big as a walnut. Some of the fruit are all red, fome all yellow, and fome mixed with both red and yellow, and others all white, of a very pleafant tafte in general; but there is a great variety, as fome more tha p or tart, fome like the tafte of cherries, others very rough like unripe apples, but moft of them fweet and pleafant, and generally goes off with a reftringency or flipticity upon the tongue, which proceeds from its tough fibres that run longway through the fruit; when cut with a knife, it turns it as black as ink. There are fome of the fruit bigger than others, but the generality of them are as big and much of the fhape of French pippins, and make an excellent cyder or wine. I, having a large orchard of about three
hundred

hundred trees, after the market was glutted with them, diftilled a fpirit from them far exceeding arrack, rum, or brandy, of which they made an admirable punch, that would provoke urine powerfully. The flowers are very fmall, and grow in tufts, of a carnation colour, and very odoriferous. The leaves much refemble the Englifh walnut-tree leaves in fhape and fmell, and are as effectual in old ulcers, cleanfing and healing them, being decocted, and the ulcers wafhed with it.

The nut hath a very cauftic oil, lodged in little partitions betwixt the two outward coats, which will flame violently when put in the fire. This oil cures the *herpes*, cancerous and malignant ulcers abounding with rotten flefh; it alfo kills worms in ulcers and chigoes; it takes away freckles and liver fpots, but it draws bliflers, therefore muft be cautioufly made ufe of; and fome make iffues with them; it alfo takes away corns, but you muft have a very good defenfive round the corn, to prevent inflaming the part. The infide kernel is very pleafant to eat, when young and before the fruit is come to its ripenefs, exceeding any walnut; and, when older and drier, roafted, they eat very pleafant, exceeding piftachia-nuts or almonds, and, ground up with cacao, make an excellent chocolate. The gum of this tree is very white and tranfparent like glafs. It hath been obferved, that poor dropfical flaves that have had the liberty to go into a cafhew-walk, and eat what cafhews they pleafe, and of the roafted nuts, have been recovered.

These trees are of a quick growth: I have planted the nut, and the young trees have produced fruit in two years time, and will keep bearing once a-year for forty or fifty years, nay, a hundred, by what I can underfland, if no accident attends them. Many are now flourifhing in Jamaica that were planted when the Spa-

C

niards had it in poffeffion; for the wood is excellent
ftrong and lafting timber.

CASSADA

Is well known in Jamaica. The root of this plant
makes a very good and wholefome bread, notwithftand-
ing the juice is a deadly poifon, called *manipucra*,
wherefore great care is taken to prefs out all its juice;
and then, dried in the fun, beat, and finely fifted,
and baked upon a flat broad round iron, commonly
called a baking-ftone, they make the cakes as broad
as a hat, which, buttered while hot, eat like an oat-
cake. I have feen feveral bad accidents happen to
negroes newly come to Jamaica, and ftrangers to
the root, who have eat of it only roafted with its juice,
which hath poifoned them: The fymptoms are, firft,
a pain and ficknefs of the ftomach, a fwelling of the
whole abdomen, then violent vomiting and purging,
giddinefs of the head, then a coldnefs and fhaking,
dimnefs of fight, fwoonings, and death, and all in a
few hours. The expreffed juice of the root is very
fweet to the palate, but foon putrifies and breeds
worms, called *topuea*, which are a violent poifon, and
which Indians too well know the ufe of: They dry
thefe worms or maggots, and powder them; which
powder, in a little quantity, they put under their
thumb-nail, and, after they drink to thofe they intend
to poifon, they put their thumb upon the bowl, and
fo cunningly convey the poifon; wherefore, when we
fee a negro with a long thumb-nail, he is to be mif-
trufted. The only and quickeft remedy for caffada-
poifon is, firft to give a vomit of ipecacuana, and then
the juice or powder of *nhambi*, which I fhall mention
hereafter. Caffada bread, milk, and fweet oil, make
an admirable poultice to ripen and break any fwelling.

There

There is a fort of caffada which is called fweet, for it may be eaten raw, or roafted like a potatoe, without any manner of prejudice or hurt, being very nourifhing, and makes a very fine white flour ; this bears a large berry.

There is another plant, called wild caffada, and is known by no other name by the people in Jamaica, But for what reafon I cannot tell, it being in no refpect like the other caffada ; they grow wild in every favanna. In the months of March and April, there is found, in the infide pith of the foot-ftalk, a hard knotty excrefcence, of an oval fhape, hard and yellowifh, of divers fizes, as from a hazel-nut to a hen's egg : I never could find what ufe they are of ; only I have obferved the boys will powder them and give it for fnuff, which will burn and tickle the nofe, and caufe greater fneezing than white hellebore. I am apt to believe they will purge violently ; for the young tops of this plant, boiled and buttered, are often given in the dry belly-ache, as alfo in clyfters, purging violently when nothing elfe would go through the patient. The feeds are like a fmall *ricinus ;* and, if they are not the true granadilla, yet they purge as ftrongly ; for two or three feeds, given by themfelves, or mixed with pills, quicken the purging quality. I knew a practitioner who always made up pill *ex duobus* with addition of thefe feeds, which made the pill work ftronger and quicker, and kept it always moift. You make the pill thus : *Take wild caffada-feeds hufked, three ounces ; cambogia, coliquintida, and fcammony, of each one ounce ; make a pill according to art ; the dofe is two or three fmall ones.* They will purge very brifkly all watery humours.

CASSIA

CASSIA FISTULA.

There are two forts that I know growing in America, whofe trees are very large, with winged leaves, four or five ftanding on each fide of the ftalk, like Englifh afh, long and fharp-pointed; the flowers are yellow and large, with five leaves with thrums in the middle, fmelling very fweet; one thrum, which is the ftyle, is longer than the reft and crooked, and is fixed to the pod as it grows. The pods differ much as to their length; *viz.* from twelve inches to eighteen; I have feen fome above thirty inches long. It gently purges.

The fecond fort is called horfe-caffia: The leaves of this fort differ extremely from the other fort, being fmall foft leaves, ftanding on each fide of the ftalk, to the number of fourteen or fixteen of a fide, of a pale green on the upper fide, and of a yellowifh green underneath, and of the bignefs and fhape of fenna, but a little more rounding. The ends of the branches, for two or three feet long, are fet full of beautiful flowers, very odoriferous, of the colour of peach-bloffoms, and very much refembling them. The fruit is much larger than the other fort, and of a very rank ftrong fmell. It hath a wonderful power to move the monthly purgations in women.

There is alfo a fhrubby caffia: It hath a fmall long pod, about the length of calavances, which is full of feeds fticking in a fweet clammy pulp, which the boys in America fuck, and which generally purges them. It powerfully provokes the terms in women.

CEDAR.

There are two forts of cedar-trees grow in Jamaica. The one fort Sir H. Sloane calls *pruno forte affinis ar-*

ber

bor maxima materie rubro laxo odorato : Thefe grow in plenty in the mountains, and, where they grow, they reckon the ground rich; they are next in bignefs to the cotton-tree that they make canoes or boats of. I have feen fome cedar-trees three feet in diameter, with nine feet in circumference. The leaves are like thofe of the common plumb-tree of America, almoft like the Englifh afh-leaves, and they have a round berry which the birds eat; the wood is foft like deal, but reddifh, having a very pleafant fmell; its gum is like gum arabic, very tranfparent, and eafily diffolves in water, wherefore the fhoemakers ufe it as gum arabic.

The other fort is called juniper cedar, and is the fame fort that grows in Bermudas : This hath leaves like the favine or fir, or pine trees; its wood is whiter than the other, fmelling more like juniper berries; the gum refifts putrefaction, and kills worms.

CELANDINE.

I have often met with this plant, and wondered how they came to call it celandine, it differing fo much from the Englifh fort; for this generally grows fix or feven feet high, with a very thick ftalk covered with a white fmooth bark, branching with a great many large leaves, and deeply divided at the ends, of a yellowifh-green colour on the upper fide, and whitifh underneath; on the top comes out a branch of a foot long, full of bunches of flowers, each ftanding on a fhort foot-ftalk, and hath in it many ftamina or threads of a yellow colour, and feed-veffels of an oval fhape, in the middle of which is a fmall brown oblong feed : All parts of this yield, in breaking, a yellow juice, like common celandine, from which it hath its name, as I fuppofe. Hernandez calls it *quauhchilli,* five *Chilli* fpecies, from its fharpnefs like Indian pepper, and

C 3   faith

faith it was planted by the Indian kings in their gardens. It is much ftronger than Englifh celandine, being very hot and drying. The juice cures tetters and ring-worms, and takes off warts and films of the eyes; but I fhould not care for ufing it to the eye, being fo very fharp.

Centaury.

There are two or three forts of centaury grow in America. One is called *cachin lagua;* a fmall fort of centaury, more bitter than the European, and confequently more full of falt; it is reckoned an excellent febrifuge.

Another fort, that grows about Panama, they make a tea of, which is aperitive and fudorific; it fortifies the ftomach and kills worms, cures intermitting fevers and the jaundice; it is alfo given with very good fuccefs in rheumatifms, &c. They take it as hot as they can, in bed, covering themfelves clofe to provoke fweat. This plant fmells like natural balfam; and is fo great a fweetener of the blood, that it is a fpecific in pleurifies and fevers, and is ufed inftead of the jefuits bark. It is found plentifully about Panama, and divers other places. That which grows in the mountains is efteemed the beft.

Cerasee and Cucumis

Is the name that negroes and fome others give to a plant growing in great plenty in Jamaica. Its fruit is much like a cucumber, and as big; therefore Sir H. Sloane calls them *cucumis puniceus,* I fuppofe from its deep-red colour, but the leaves are much fmaller, jagged, and divided; the fruit generally of the fize of a lemon, of a yellowifh red without-fide, with blunt tubercles; the infide is of a moft glorious red colour,

having

having feveral large red feeds, in bignefs and fhape of tamarind ftones or feeds. I have obferved, if you put the point of the fmalleft pin or needle into any part of the fruit, it will all fly open in quarters, or many parts, turning, as it were, the infide outward, with a fort of guft or explofion, or as if it were fenfibly touched. Some make fine arbours with this plant, it always climbing to any thing it is near, growing fo thick you can hardly fee through it. Some fuck the feeds, having a fweet red pulp about them; but the fruit is very hollow, like pops, and purges excellently well. The negroes cure the belly-ache, by mixing with it Guinea pepper. Both leaves and fruit are a great vulnerary: A decoction or infufion of the roots in water, wine, or broth, wonderfully evacuates watery humours, and prevails againft the yellow jaundice, obftructions of the liver, fpleen, bowels, and mefentery. The root, powdered and given with cream of tartar (from a fcruple to forty grains), doth the fame; a fyrup of the fruit doth the like. The diftilled water from the leaves and fruit, mixed with *fal nitri*, makes a beautiful wafh, and is good againft the St. Anthony's fire, or any rednefs of the face; inwardly given, with loaf-fugar, it cools and abates the heat of fevers. The oil from the fruit cures burns, and takes away fcars. The wild cucumber grows in great plenty in moft parts of America, from the juice of which I have made *elaterium*. We have of the common garden cucumbers, as good as in any part of the world.

CHERRY-TREE.

There are two or three forts of what they call cherry-trees, but not to compare with thofe of England. The clammy cherry is a beautiful tree to look at, and bears a fine red fmall round cherry, but it is clammy in the

mouth,

mouth, not fit to eat; but birds delight to eat them, and turkies and fowls will devour them. The Barbadoes cherry is of a very pleafant tartnefs, and makes an excellent red jelly, which allays the heat of fevers. The Brazilians call them *ibipitanga*.

CHILI CARDINAL FLOWER.

This is called in Chili *tupa*. Its flower is red, and they grow generally on mountains. The root and bark yield a venomous milk, which will endanger the eyes like fpurge. It is faid, that the very fmell of the flowers caufes vomiting, and the whole plant is reckoned a violent poifon.

CHINA-ROOT.

This root grows in great plenty in America. It hath a root as big as one's arm, is crooked and jointed, with knobs at every joint like fome canes, very tough, and when young of a green colour, very full of prickles like a rofe bufh or brier, but when older has little or no prickles, and will grow to be bigger than a man's thumb, and fometimes ten or fifteen feet high. The leaves are like the *fmilax afpera*, or farfaparilla; they are cordated, fmooth, of a very dark-green, with nerves like thofe of the Englifh plantane-leaf. At the end and between the twigs come out the flowers, feveral together, but from one centre, ftanding on an half-inch piftil, of an umbel fafhion; each hath fix petals, with very fmall green apices, ftanding round a green fhort ftylus; after, follow fo many blackifh berries, round, and of the bignefs of thofe of ivy, having an unfavoury purple pulp, with a purple ftone as big as that of the haw. Sometimes a gum is to be found, which the Indians call *tzitili*, which they chew to ftrengthen or faften their teeth. I have feen a fort

much

much whiter, without and within, than the common
fort. The ufe and virtues of this root are fo well·
known for and in venereal cafes, as I need not give any
further defcription of it; only juft mention what ufe
Dr. Trapham made of it in fuch cafes, who practifed
many years in Jamaica; but he firft gave the following
electuary:

*Take pulp of tamarinds and caffia fiftula, of each
half a pound; juice of femper vive, three pounds;
fmall red pepper or capficum, dried, one fcruple; Winter's cinnamon, one fcruple and an half; of melaffes,
clarified with the white of an egg, a pound and an half.
Put all thefe into an earthen pot, which place in the fun,
ftirring the mixture with a wooden fpatula, two or three
times a-day; let it ftand till it thickens to a due confiftence of a foft electuary, which keep for ufe as a general
purge.* The dofe, from half an ounce to an ounce
and an half; in clyfters, two ounces. Let the patient
take half an ounce of this, or two good broad knife-
points full, in the morning fafting, and as much at
night going to bed, two hours after having eaten fome
fpare fupper; continue every other or third day till the
gonorrhœa ceafes. The dofe may be leffened according
as it works; and thofe days they do not purge at night,
let them take a drachm of china-root in powder,
drinking the following decoction or infufion of china-
root, warm, to fweat with; the drink ought to be
made new every day, without being fermented with
fugar or age. The water is only to be boiled as that
for tea; then fo much china-root, fliced, added thereto
as may make it of a claret colour; there can be no ex-
cefs in the root, neither need there to be added, fave for
palate fake, a little fugar, for it is better without; let
him drink thereof every night in bed plentifully, about
two quarts, the better to fweeten the four juices, which

china-

china-root powerfully doth in thefe cafes as well as in others, fuch as gouts, tertians, hectics, confumptions, &c. and then, to complete the cure and ftrengthen the fpermatic veffels, let them take hog-gum in pills for fome time.

I am very well affured, that this Weft-India china-root is in every refpect as efficacious and as valuable as that from the Eaft-Indies; but the great difficulty is how to preferve it from the worms; for, in a month or two, it will be bored, and all the farina or mealy part fcooped out, by a large white maggot with a red head, that breeds in it. I have tried feveral ways to prevent it; the only way was, to trim it well of all its foft knobs, and then to bury it in white lime.

CINNAMON.

We have only one fort, called Winter's cinnamon, from one captain Winter, that firft carried it to England, where it is well known. The bark hath a fmell refembling the common cinnamon, but much hotter and whiter; that taken from the branches is better than that from the body of the tree. It hath a laurel-like leaf, much like the pimenta; its fruit is a little berry, which is violent hot, and much like *cubebs*.

See more of it under Winter's Bark.

CITRONS,

Both fweet and four, we have in great plenty, as large and as good as any in the world.

CLARY.

Befides the garden clary, we have a very common plant, that grows every where in Jamaica, called wild clary. The ftalk is large, green, and hairy, rifing about two feet high; the leaf like garden clary, hav-

ing

ing many five-leaved flowers, of a pale-blue colour, set in a double row on the upper fide of the branches, and turned like a fcorpion's tail. Like the *heliotropes*, it cleanfeth and confolidates wounds and ulcers, and is good againft inflammations of the fkin. It is boiled with cocoa-nut oil, to cure the fting of fcorpions and the bite of a mad dog.

CLOVE-STRIFE.

Two forts of clove-ftrife grow in America; firft, the broad fort, which Fuillee calls *onagra laurifolia flore amplo pentapetalo;* the fecond fort is the female or leffer, called *onagra minor flore luteo pentapetalo.* The Indians highly efteem thefe two fhrubs, making a poultice of the leaf, which mollifies and diffolves all kinds of tumours, which are very common in thefe parts. They delight to grow by river-fides.

COCA.

This herb is famous in the hiftories of Peru, the Indians fancying it adds much to their ftrength; others affirm, that they ufe it for charms; as for inftance, when the mine or ore is hard to work, they throw upon it a handful of this herb chewed, and immediately get out the faid ore with more eafe and in greater quantity, as they fancy. Fifhermen alfo put fome of this herb chewed to their hook, when they can take no fifh, and they are faid to have better fuccefs thereupon. In fhort, they apply it to fo many ufes, moft of them bad, that the Spaniards prohibit the ufe of it; for they believe it hath none of thofe effects, but that what they attribute to it is done by the compact the Indians have with the devil. The leaf is a little fmooth, and lefs nervous than that of the pear-tree; the fhrub does not grow above four or five feet high.

The

The greateſt quantity grows about thirty leagues from Cicacia, among the Yunnas, on the frontiers of the Yunghos. The taſte of it is ſo harſh, that it fleas the tongues of ſuch as are not uſed to it; it occaſions the ſpitting of a loathſome froth, and makes the Indians who chew it continually ſtink abominably. It is ſaid to ſupply the want of food, and that, by the help of it, a man may live ſeveral days without eating, and not be ſenſibly weakened. It is thought to faſten the teeth, and take away their diſtempers; and it anſwers in all reſpects the purpoſes of tobacco,

Cocoons

Is a great large broad flat bean, reddiſh, and hard when dry, and round, fit to make ſnuff-boxes of, and may be poliſhed very fine. The inſide kernel is very bitter, and vomits and purges ſtrongly. Piſo tribes it among his poiſon plants. They grow only in the mountains, and run up upon the higheſt trees, with ſtalks as big as a man's wriſt; and have a broad crooked pod, about twelve or fifteen inches long and ſix inches broad, firſt green, and then black when ripe.

Colilu or Culilu.

This plant is more for food than phyſic, and is much the ſame as Engliſh ſpinage; ſome ſay it exceeds it, eſpecially young and freſh gathered. It grows in great plenty every where, without cultivating, after rains; and is of great ſervice to poor ſlaves, who, if they can but get ſalt to ſeaſon it (otherwiſe it is apt to purge them, if they eat too much of it), they will live upon it weeks together.

Contrayerva.

This is ſo called in Jamaica from its great efficacy
againſt

againſt poiſons, but is in no reſpect like the Spaniſh contrayerva; for this plant hath a long round geniculated root, in ſhape and bigneſs of long birthwort; ſo are its leaf and flower. It hath a round green climbing ſtem, taking hold of any tree or ſhrub, riſing ſix or twelve feet high, covering them with its numerous branches. The leaves ſtand on the main ſtalks, cordated, of a dark-green colour; the flowers ſtand on a three-inch foot-ſtalk, like other birthworts, of a yellowiſh colour, the lip covered with a purple farina; the fruit is hexangular, two or three inches long, containing ſix cells, full of ſmall flat odoriferous yellowiſh-brown ſeeds, of the ſhape of an heart. The roots and ſeeds are very bitter, hot, and odoriferous, and are moſt excellent alexipharmics or counter-poiſons, ſtrengthening the heart, ſtomach, and brain; they cure the bites of ſerpents, and the poiſon of Indian arrows. I am of opinion, it exceeds the Spaniſh contrayerva, eſpecially in dropſies. I have ſeen wonders done with it: It drives out the ſmall-pox, meaſles, ſpotted fevers, plague, or any malignant diſtemper; it gently purges ſome by ſtool, but never fails working powerfully by urine, and ſometimes by ſweat. I have known it recover ſeveral in lingering diſtempers, when their appetites have been wholly loſt and the uſe of their limbs, and that only by drinking a ſimple decoction of the root in water; but in wine it makes the beſt ſtomachic, it being exceſſive bitter and aromatic; alſo this makes the beſt bitter wine in the world, exceeding all in the diſpenſatories, or Stoughton's drops; and, if you add ſteel to it, it cures the green ſickneſs, dropſies, opens all obſtructions, ſweetens the blood, and reſtores it to its due craſis.

COOPERS

Coopers Withe.

This withy plant is so called because coopers make hoops of its stalks or withes, being very tough and flexible; and although this plant doth not climb or twist round other plants, yet it cannot support itself, but, growing by the side of any tree, it leans upon it, and, by its many branches, will overspread it. It hath a leaf of the breadth and shape of laurel leaf, but not so thick or glossy; its flowers are inodorous, mixed with purple streaks; and then follow small round berries, growing all along the spikes or twigs of the shrub, in colour, shape, and bigness of elder berries, for which reason some call it Spanish elder; but that is another plant. It hath an uncommon excrescence, that is found growing among the branches at one time of the year, which is in shape exactly like the stomach of a man, having a thin membrane or skin over it, interwoven variously with innumerable small reddish veins; it adheres to a tender soft stalk, which runs through the upper part of the excrescence. This plant is of divers physical uses. Bess Walker, who kept a tavern in Port-Royal in Jamaica, before the great earthquake in 1692, used to make a famous drink, reckoned of use in venereal cases; for which she boiled the young tender withe sliced in water with a little *lignum vitæ* bark, worked it up with sugar or melasses, and then bottled it; it drank brisk like bottled ale, only bitterish. It is a good stomachic, and opens obstructions. The Indians make a bath of this plant; they strip naked, and place themselves so as to receive the fumes or steam of the liquor, being covered all over with a blanket or pavilion, after which they are put to bed, and rubbed very well; by this method, they recover the use of their weak and numbed limbs, and comfort their bowels.

COPAL.

COPAL.

This gum flows from the trunks and branches of several large trees growing on the mountains in America, with fruit like our cucumbers, but of a dark-grey colour, in which is a mealy flour, of a very good taste. It is a fine clear pure transparent yellowish-white gum, very odoriferous. This and *gum animi* are much the same.

CORALS *and* CORALLINES.

I do not see writers of America take so much notice of the coral kind as Sir H. Sloane, who, in his Natural History of Jamaica, makes six sorts of white coral, and seven sorts of bastard corals, or corallines. I never saw any red in Jamaica or America, but I have often seen red worts sticking to the white coral, as big as pease. The white coral is so plentiful in Jamaica, that they burn it, and make a very white lime for building. I am of opinion, that the white is every way as medicinal as the red; the corallines are said to kill worms, but it doth not stand to reason.

COTTON.

And, first, of the useful shrub that so much cloth is made of, although it is but a shrub, that seldom rises above ten or twelve feet. Its large leaves have five points, in shape of the English maple or sycamore; the smaller leaves, nearest the fruit, have only three points, of a deep-green. The flowers are like the tree-mallow, or holyhock, but not so open, of a yellow colour; they are supported with a foot-stalk and green cup, composed of three triangular jagged leaves, which inclose them but very imperfectly; they are yellow at the top, and streaked with red below. The flower or blossom is succeeded by a green fruit like a rose-bud, which,

which, when full ripe, grows as big as a little egg, and divides into three or four cells, each of them filled up with between eight and twelve feeds, almost as big as peafe; thefe are wrapped up in a woolly fubftance, known by the name of cotton, which fticks to the feed, and, as the pod opens, they drop down together, if riot gathered in time. This cotton fhrub differs much from that which they cultivate at Malta, and many other places in the Straits, and throughout the Levant, which is only a very little annual plant, that is fown and renewed every year: Now this American cotton will ftand many years; only pruning them after they have done bearing, and they bear for feven years to-gether. The cotton is fit to gather in January, and may keep picking till May, and then you fhould trim them, or plant anew. To part the feeds from the cot-ton, they have a little inftrument, called a gin, with two rollers as thick as a finger, and, with two wheels turning contrary ways, pinching the cotton, and drawing it through between the two rollers, leaving the feed be-hind. Thefe feeds are faid to be good againft the bloody flux, and are counted pectoral; the oil clears the fkin of fpots and freckles.

Cotton-Tree.

We have two forts of large cotton-trees in America; one whofe wood is more red, the other very white, and bears a fruit as big as a large cucumber; which, at a certain time of the year, opens, and lets drop a fine down or filk cotton, which, with induftry, may be found to be of great ufe, I believe in hats, in the room of beaver; but at prefent of no known ufe among the Spaniards or Englifh. There are fome of thefe trees fo large as to make a canoe that will carry fifteen or twenty hogfheads of fugar, each hogfhead weighing

from

from 1 2 to 1 6 *cwt.* The feeds of thefe trees are much of the fame quality as the other cotton; its green bark, made into a poultice, is good againſt inflammations, and confolidates fractured bones.

COWHAGE, *or* COWITCH.

This plant is tribed amongſt the *phafeoli*. It is fo well known, that there needs no defcription; and may be felt when not feen. The root provokes urine, the bean the fame, and, boiled in oil, eafes the gout and St. Anthony's fire.

CURRANT-TREE.

This is fo called becaufe its fruit is of the ſhape and bignefs of a currant. It is a moſt ſtately tall tree, as big as the Engliſh elm, and is always green, having a laurel leaf, and a whitiſh ſweet flower in bunches; after which come bunches of fmall yellowiſh round berries, in taſte like the Engliſh haw. The birds delight to eat them, and build their neſts in the trees.

CURRATO.

I take this name to be a corruption of *caragua*, for fo it is called in Brazil. It is of the aloetic kind, and I have made an extract out of it much like aloes. The leaves are very large and fucculent, five or fix feet long, and but five or fix inches broad, having a black ſharp prickle at the end, and fmall hooked prickles upon the edges. It hath, about April, a ſtalk as big as a man's leg, about fixteen or twenty feet, running up, as fome affirm, in a night's time; but that I quef- tion. The top of it, in May, is garniſhed with fine yellow flowers for three feet down: The people in Jamaica gather them for May-poles. After the flow- ers, appear many pods, bigger than a man's thumb,

D   full

full of thin membranaceous feeds, like parfnip feed. The juice of this plant fcours like foap, and in fome cafes is ufed as fuch. It is full of long and ftrong fibres, which they feparate as they do the filk grafs, and make lines and ropes of it. Its juice, with a little fugar, will powerfully force the terms, is a great diuretic, and forces gravel or ftone; the leaf, roafted in the fire, takes away the pain and weaknefs of the limbs. The extract eafes the pain of the gout, and ftrengthens the part, if ftrained, ftuck upon leather, and applied thereto: At firft applying it feems to increafe the pain, for it draws ftrongly a fort of dew from the part, but, after three or four hours, the pain ceafes, and the part grows ftronger every day; it muft lie on until it drops off. I always ftuck it upon white paper, and applied it to myfelf, and in two or three days was able to walk three or four miles, &c. If the extract is not well boiled, it will draw pimples, and caufe a great itching. I have given it inwardly in pills, with good fuccefs. It is alfo called *maguly.*

CUSTARD-APPLE,

So called becaufe the pulp is foft when ripe, white, and fweet like a cuftard. The leaf of the tree is in fhape of the peach-tree leaf; its fruit is of the fhape of four-fop, but not quite fo large, and of a brown-ruffet and yellowifh colour, and a rednefs on one fide when ripe.

There is another fort called water-apples, becaufe they are found growing along the banks of rivers. I have feen fome of the apples bigger than what they call a cuftard-apple in England; they are of a cold watery quality, to that degree that they are reckoned a fort of poifon. The alligators live chiefly upon them, one time of the year, watching their falling into the

water;

water; there are great quantities of thefe trees growing by the fides of the Rio-Cobre, near the lagoons.
See Water-Apples.

DAISY.

. We have a fort of daify grows in Jamaica. It is a
great vulnerary, and helps all difeafes of the lungs;
its juice cures confumptions of them.

DANDELION.

It is fingular againft obftructions of the *vifcera*. The
juice of the leaves and roots, given in Madeira wine,
purifies the blood and juices of the body, and provokes the *catamenia*. The diftilled water, made fharp
with oil of vitriol or fulphur, prevails againft fpotted
fevers and plagues.

DILDOES

Is the name of a plant which grows in all the fouth-
ern parts of America, and in Jamaica. Some merry
perfon gave it the name of dildo; but in other places
it is called flambeau, torch-wood, or prickle-candle,
it being in the fhape of four candles joined together in
angles, growing one out of another, like the *raque*,
and are from eight to fifteen feet long, fet with fharp
prickles all round from top to bottom, green, and full
of juice. Some bear a yellow fruit, others blood-red,
without-fide, but of the fame colour as the reft within;
which is a white fweet pulp, full of fmall black feeds;
and they have all a large white flower, fmelling very
fweet, which always comes out of that fide of the plant
next a fouth fun. Its fruit is as big as large apples.

When they grow old, and the green juice dries
away, there is a yellow hufk, or fhelly fubftance, appears full of holes like net-work, which is called torch-

D 2   wood,

wood, for it will burn like a candle and torch; and I have known the Indians fill the hollowness of thefe with a bituminous fubftance, making fine flambeaux.

DODDER

Is a ftrange fort of a plant, running over and de-ftroying every plant it comes near, therefore is called by fome hell-weed, or devil's guts. It hath ftrong yellow filaments, by which it ftretches over very large and high trees, covering the plant which it feeds on, and deftroying it. The flowers are white and conglo-merated; it hath a pale-coloured feed, fomewhat flat, and twice as big as poppy-feed. This devouring weed generally takes after the quality and properties of the plant on which it grows; but in general it hath a ca-thartic quality, and opens obftructions, &c.

DOGSBANE.

There is great variety of dogfbanes in America; and they will be mentioned, as they come, under other names. The blood-flower, mentioned already, is one fort.

DOG STONES.

There are two forts of dogs ftones grow in the fa-vannas in America, with double tuberous roots, much like thofe in England. It hath a fpecific quality to help impotency in men or women; and the effence, juice, or extract, taken morning and night, in a glafs of generous wine, is faid to poffefs prolific virtues.

DOG-WOOD.

This tree is fo well known in Jamaica, that it need-eth no defcription, being the chief and moft lafting timber in America, every way as good as the Englifh oak,

oak, and having much fuch a leaf; but they never grow fo large. Its bark hath a very ftrong rank fmell, and poifons fifh. It makes a glorious fhow when in bloffom, which it will be when there is not a green leaf upon it: The bloffoms are very white and fweet, fmall, and in bunches as full as the tree can hold; afterwards come bunches of a membranous fubftance, looking like hops at a diftance; in which is contained its feed. The bark is very reftringent: I have made a decoction of this bark, which would cleanfe and ftop the great flux of ulcers, and make them fit to heal, and cure the mange in dogs.

DRAGON'S BLOOD.

The trees that affoid this gum grow in both the Indies and in Africa. Indian dragon's blood is a gum that diftils or drops from the trunk of a tree, whofe leaves are like fword-blades, of half a foot long, and green; at the bottom of which grows round fruit, of the fize of Englifh cherries, yellow at firft, afterwards red, and of a beautiful blue when ripe; from which, having taken off the firft or outward fkin, it appears like a fort of dragon, which gave it this name. They cut the trunks of the trees, and there prefently flows a liquid liquor, that is as red as blood, which hardens and forms itfelf into little brittle tears or crumbs. When the firft fort is fallen, there drops another fort, which is brought us wrapped up in its own leaves: Chufe that in little tears, clear and brittle.

Dragon's blood of the Canaries flows from the trunk and large branches of two different trees; one of which has leaves like a pear-tree, but longer, and the flower refembles tags at the ends of laces, of a very fine red. The leaves of the other come nearer to the cherry-tree, and the fruit is yellow on the edges, of the big-

nefs

nefs of a hen's egg; in which is a nut of the fhape of a nutmeg, containing a kernel of the fame figure and colour: Thefe trees grow plentifully in the Canary iflands. I faw one in Madeira, in a Portuguefe garden, in the year 1696; and in the little ifland of Poito Sancto they grow in great plenty. They grow alfo in great plenty in America, efpecially about Portus Sanctus, and are there called *kinnabris*. The juice or gum they make into balls: The beft fort is in drops; a coarfer fort is in round cakes; and the coarfeft of all is that in great lumps. It is counterfeited with fenegal, and coloured with a tincture of brazil.

Duck's Meat, or Pond-Weed,

Grows in ftanding waters, and is accounted cold and moift, taking away inflammations of the liver after a peculiar manner. It is reckoned good in all outward hot fwellings, or difeafes of the fkin, and all inflammations; ftops fluxes of blood and ruptures.

Dumb-Cane.

This is fo called becaufe, if any body bites of it, they cannot fpeak for fome time; for it burns and benumbs the tongue, and caufes a great flux of fpittle. It grows in joints, appearing like green fugar-canes, and therefore fo called; and fome have been deceived in taking them for fugar-canes. Its fruit is like fome of the *arums;* but the leaves are like Indian fhot, or our water-pepper. It hath been ufed with good fuccefs in the dropfy, in the following manner: *Take the greenefs and moft juicy, and beat it in a mortar into a kind of pulp; then add thereto double the quantity of hog's fat, or rather tortoife fat, or fnake's fat; the which having agitated ftrongly together, let it lie for fome days; then beat it well again, and keep it for ufe;*
observing,

obferving, the longer it is kept it better anfwers the intention; but it muft be heated and beaten now and then, left worms breed in it. To prevent that, and alfo render the ointment more fine, fafe, and agreeable, take of the mafs, beaten as before, warm it, and ftrain it through a coarfe cloth, which boil up to a due confiftence, and keep for ufe; which is thus: Take of this ointment, and chafe it warm into the fwollen parts, and apply as a cataplafm to the *fcrotum;* by which method the watery humour will be difcharged.

DWARF-ELDER.

This plant is fo called becaufe it fomething refembles the European dwarf-elder, being a fhort plant; with a round jointed ftalk and a reddifh fruit; but its leaves are much like the large Englifh nettle, with large nerves or veins. It delights in fhady places. A colonel, who had lived many years in Jamaica, affirmed to me, that it was a certain cure for the dropfy, purging off the water gently by urine and ftools, by giving its juice or ftrong decoction.

DYING PLANTS.

Befides the medicinal plants, they have others for dying in fuch manner that the colour will not come out with often wafhing with foap. Such is the root of *riclbon,* or *raize-rue,* a fort of madder, the leaf whereof is fmaller than ours; and of which they boil the root in water to dye red. *Poquell* is a fort of gold-button, or female fouthern-wood, with green checquered leaves, which dyes yellow, and holds well; the ftem of it dyes green. *Indigo,* that dyes blue, and the *blueberry tree. Panke,* or *panque,* dyes black, and tanners boil the roots of it with their fkins, which very much thickens them: *Opoponax* doth the fame. *Itin*

wood;

wood; they ufe the chips, with the roots of *panke,* mixed with a black earth called *robbo;* thofe they boil in water, with which they dye their ftuffs of a fine black, which fades not like thofe of Europe: Befides *fuftic, logwood, brazil, braziletto, nicaragua, prickly pear,* and feveral others. *Docadilla* is a fort of ftone fern, which grows in great plenty in South America; the decoction of which, when drank, refrefhes after the fatigue of travelling, and is efteemed a great fweetener of the blood.

EBONY.

This tree grows every where in the favannas. It hath a fmall leaf like box, and a yellow flower like Englifh broom, and, after rains, puts forth its flowers, making the favannas look like Englifh broom-fields. Its heart, or inner part, is as black as jet. The oil of it cures the tooth-ache, cotton being dipped in it, and put into the hollownefs.

EDDOS.

Both the leaf and root of this plant are good food. There are three or four forts of them: Pifo calls them *ta acba,* and from thence came the word *tayas;* but they are the largeft fort, and apt to caufe a heat in the throat, which they call fcratching the throat, fo that only negroes and hogs eat them; and they muft be well boiled to correct that peccant juice, which is like what is in *aramithier.* The leaves, efpecially of the leffer fort, are very much like *dragon arum,* or wake-robin.

ELDER.

There is a fhrub in Jamaica known by the name of **Spanifh** elder, having a green jointed ftem, full of
pith,

pith, like Englifh elder, but hath none of ts fruit, but a fort of a *julus*, like the long pepper. It cures the cholic, taken in powder from a drachm to two drachms, in Madeira wine. A certain gentleman af-firmed to me, that he gave it to one of his negroes that had a venereal taint, which brought away fuch a quantity of flime and urine, it was incredible, and cured him. A bath of the whole plant recovers weak limbs to their ftrength.

ELEMI.

This is a white refin gum, inclining when new to be greenifh. It flows from the trunk of a tree, of a moderate height, whofe leaves are long and narrow, of a whitifh green, filvered on both fides; it has a red flower, that rifes from a little cup of the colour of the leaves; and the fruit is of the colour and fhape of olives, from which they are called wild olives. Chufe that which is dry, neverthelefs of a white colour, foft, tending to green, the fmell fweet and pleafant, and will readily flame. It is rolled up in America in leaves of the palmeto, which fome call thatch and flags, that they make brooms of: Take care it be not factitious, which is made of turpentine and oil of fpike, and is called *gallipot*. There is a large tree found in America, whofe wood is white, and the leaves like thofe of the bay, but a great deal larger, which affords abundance of gum called *cleban*, and is like the true *gum elemi*.

ELM.

We have a tree in Jamaica called Spanifh elm, which hath a very fweet pleafant fmell, almoft like a rofe. This tree is very common, and known to moft inha-biters in Jamaica. The coopers make hoops of the

young

young ones for fugar hogfheads. The heart of it is a very fine veiny wood, and would be of great ufe to joiners for cabinets. The oil is not inferior to *rhodi-um*, having the fame ufe and virtues.

ERYNGIUM, *or* FRINGO, *or* SEA-HOLLY.

It is alfo called *eryngium Americanum fœtidum*. It hath fix or feven round fmooth whitifh roots, going ftraight down into the earth, fometimes ten inches long, which uniting towards the furface of the earth, fend forth leaves, fpreading on the ground on every hand, five or fix inches long and one broad, deeply ftriated or jagged, with little foft prickles on the edges ; the tops or heads are like other *eryngiums*, having feveral brown feeds fet round a fmall column. This plant hath a very penetrating ftrong fmell. In Surinam it is called *itubu*, or *fuga ferpentum*, becaufe ferpents will not come where it grows. It is excellent againft the bite of ferpents, a great ftomachic, expels wind and cafes the cholic, provokes urine and the *catamenia*. It taftes like fkirrets; and, having a ftrong fmell, is good againft hyfterics, and that when only fmelt to.

FEMALE FERN

Is fomewhat bitter, with fome aftringency. Dr. Nicholas Andry, of Paris, faith, it is one of the propereft things in the world againft flat worms, and has the advantage over all other things, in that it is agreeable to all forts of people, to thofe that are in fevers as well as thofe who have none, to women with child and fuch as are not, to infants, old and young people; it allays all the fymptoms in the fick, and procures perfect eafe, fo that it may be given like nurfe's milk. The greater and leffer female ferns are known by the blacknefs of the bottom of the ftalk, but it is not fo

black

black as black maiden-hair. Dr. Andrews's great remedy againſt worms, which brought away the ſolitary broad flat worm, four ells and three inches long, is as followeth: *Take diagridium, cream of tartar, antimonium diaphoreticum, of each half a ſcruple; rhubarb, and the root of female fern, of each half a drachm, pulverized and mixed;* and let the patient take it in fat broth, at his uſual hour of riſing in the morning. This doſe may be increaſed or leſſened according to the age, ſtrength, or conſtitution.

You may with good ſucceſs give of this female fern by itſelf, mixed with honey, two drachms; or you may give three drachms of it in a glaſs of white wine; you may give half a drachm, mixed with a little honey or melaſſes, to a ſucking child.

FENNEL.

Grows plentifully in Jamaica.

FERNS.

Ferns are the greateſt tribe in Jamaica, among the vegetable kind. Sir H. Sloane makes ninety-ſix ſorts, including the hart's tongues, ſpleen-worts, maidenhairs, wall-rues, moon-ferns, and polypodies, as alſo the moſſes and capillary plants. All the ferns in general are much of the ſame nature, as drying, abſterſive, and a little reſtringent; yet as they differ in efficacy, I ſhall mention them as they come in their alphabetic turn.

FIG-TREES.

Beſides the delicious Spaniſh fig, we have a ſort of wild figs, growing ſpontaneouſly in moſt parts of Jamaica, whoſe trees are very large and ſpreading: Sir H. Sloane calls them *ficus Indica maxima,* and makes

five

five forts of them. They differ a little in fhape, big-nefs of fruit, and largenefs of leaf; but otherwife little or no difference, all having a milky juice, which is dangerous if it flies into the eyes: The juice is thick-ened, by the fun and art, into a gum like bird-lime. It is rare to fee any of thefe trees grow up ftraight of themfelves, but have generally fupporters; for, grow-ing by the fide of another, they clafp round it, and when it hath got fome height, it then puts out little branches like a withe, which grow downwards to the ground, where they take root, growing bigger and ftronger like ftilts, and then fpreading on the top, they overcome and deftroy its firft fupporter. There is both white and red, but both very foft, like deal, of which the negroes make bowls, trays, and fpoons. Its fruit is about the bignefs of an apricot. They are as large as the cotton-tree, but feldom ftraight. That which hath a reddifh wood, I am of opinion, the *balfam cativy* is got from, or at leaft a balfam may be got as good.

Fingrigo.

I believe fome negro gave the name, for it is very full of hooked prickles, like cock-fpurs; and fome call the plant fo, which is well known in Jamaica. The bloffom fmells as fweet as the Englifh May. The feeds, when dry, ftick faft to any thing they touch, like burs: I have feen ground-doves and pea-doves, that covet to eat the feeds, ftick fo faft about them that they could not make ufe of their wings, fo that you might take them up in your hands. The root of this plant negroes ufe in venereal cafes.

Flax-Weed.

All forts of flax-weeds are in fome degree, more or lefs, effectual to provoke urine and expel dropfical hu-mours;

mours; they provoke the terms, force the birth and after-birth, and are good in cancers and fiftulas. The juice, mixed with hog's fat, is excellent for the *hæmorrhoids* or piles.

Flea-Banes.

Many forts of flea-banes grow in Jamaica. They have all of them pappous feeds, or downy, like groundfel. The flowers ftand in clufters, without foot-ftalks, on the top of the plant, differing in colour; as fome fort hath a yellow flower, fome of a pale-blue, and fome purple; and, as they differ in flower, fo they differ in leaf, fome longer, and fome broader: As the *Peru chilca* is a long flea-bane, it has a pleafant fmell, the Indians make a tea of it to ftrengthen the ftomach. The *manga pak* is a round flea-bane, in Peru: This is a fweating herb, aftringent, and vulnerary; the natives drink a tea of it in cholic pains, dyfenteries, and other fluxes; it grows four feet high, with blueifh flowers, in the vallies about Lima.

Flore de Paraiso, or Flower of Paradise,

Is a very beautiful fhrub, bearing twenty or thirty flowers of different colours. They grow in the plains or favannas of Truxillo, and many other places. They have now a fort of them in England, which they call *balfamina*.

Floripondio.

The fcent of it is extraordinary fweet, efpecially in the night. Its flower is of a bell-fafhion, eight or ten inches long, and four in diameter; the leaf is downy. It is an admirable difcuffer of fwellings. In Chili it is called *datura*. It is fomething like the *ftramonium*, but its fruit is fmoother.

Flower-

FLOWER-GENTLE, or AMARANTHUS.

There are many forts of them growing in Jamaica, but are all much of the fame nature and quality, being reftringents, flopping all forts of fluxes, efpecially of blood, and overflowing of the *anus*. You may either give the feed in powder, or the juice and decoction.

FOUR O'CLOCK FLOWER.

This plant is fo called in Jamaica from its opening and flutting every four hours, night and day, as they have obferved there. I have feen of all colours. They have of this plant now in Englifh gardens, calling it *marvel of Peru;* others make it a fort of jalap. It hath a root exactly like jalap; but its fluk, leaves, flowers, and fruit, are different. I have cut thefe as they do jalap, and, when cured, he muft have a good judgment to know the difference; and I have tried to get the refin out, as of jalap; but I never could get above half an ounce out of a pound of root, whereas we commonly get an ounce and a half of refin, or two ounces, out of the true jalap. Now if the purging quality lies altogether in the refinous part of the root, then this difcovers the difference of the two roots: But I am of opinion, that all the purging quality doth not lie in the refin; for this four o'clock flower root, given in powder, works as well as the other in powder, but giving four times the quantity, and is of the fame virtue. Its fruit is black, round, and rough, without-fide; which fkin being taken off, there appears a feed as big as an Englifh pea, of the colour of Englifh wheat; and under that thin fkin is a fine white flour or meal, very foft to the touch, and taftes like wheat-flour; which I believe will purge as well as the root.

Fox-

FOX-GLOVE, *or* FOX-FINGER, *or* FINGER-WORT,

Grows in America. The Spaniards call it *acalda*, and have a proverb, that *acalda* falveth for all fores. Made into an ointment, with hog's fat and a little green tobacco, it heals all forts of wounds or ulcers; and, inwardly taken, heals ulcers of the lungs.

FRUTEX BACCIFERA, *or* CLOVEN-BERRIES.

I have feen of thefe frequently, as I have rode along the roads: I obferved the birds eat of the fruit; but its medicinal qualities I am a ftranger to. From the flowers flow out black berries, about the bignefs of fmall floes, cleaving into two for the moft part; whence the name.

FUMITERRY.

There are two forts of this plant grow in Jamaica. They open obftructions, and are therefore good in the jaundice, and alfo very beneficial in all cutaneous difeafes.

FUSTIC.

The fruit is aftringent and cooling, and makes excellent gargles for fore mouths and throats. It is faid, that the falt made out of the afhes of this wood, ten grains with treacle or mithridate, given for three or four days fucceffively, gives immediate eafe in the gout and rheumatifm. Pommet faith, no medicine is like it.

GAMBOGE.

It is alfo called *gum gutta*, or *gutta gamba*, and *gamandra*, in America *ghitta jemore*, for it grows in Peru as well as in China. It flows from a creeping prickly plant. [*See the figure of it in Pommet.*] Chufe that which is

pure,

pure, fine clofe like aloes, but of the colour of fine,
turmeric, inclining to an orange-colour, free from
rubbifh or filth in it. It is either made up in round
rol's or cakes, and eafily diffolves in water. It is a
ftrong purger of watery humours, and works upwards
and downwards; dofe from fifteen grains to thirty.

GARLIC-PEAR

Is very common in Jamaica. The fruit is cooling
and reftringent. Its leaves are an excellent remedy,
outwardly applied, to take away all inflammations of
the *anus* and *hæmorrhoids;* and alfo to eafe pains of
the head and ears.

GERMANDER, *or* WATER-GERMANDER,

Called *fcordium,* hath a fmall fibrous root, and a
tough four-fquare ftalk, lying fpreading on the ground,
three or four feet long, fending out leaves two and
two of a fide, oppofite to one another, oblong, and
without any foot-ftalk, jagged about the edges, hoary,
of a rank fmell, and fomewhat clammy; the flowers
are blue, and four-leaved; after which come black
cornered feeds. It is a fpecific or counter-poifon againft
infectious, contagious, or epidemic diftempers. It is
good againft the ftrangury, and provokes the terms.
You may take the juice, infufion, decoction, or effence, which takes away the gnawing pains of the ftomach, fides, or *pleura.* Infufed in reftringent wine, it
is good againft fluxes. The powder is given from half
a drachm to a drachm, in its diftilled water or fyrup,
to facilitate labour; it opens obftructions and kills
worms.

An excellent electuary for the lungs; *viz. Powder
of fcordium, one pound; refin, in fine powder, half a
pound; juice of creffes and honey, a fufficient quantity*

to make it into an electuary. The dofe is an ounce, morning and night. Fracaflorius invented a comfit of *fcordium,* called *diafcordium.*

GINGER.

Ginger hath a broad and flat root, with feveral divided parts, almoft like fingers. It hath many fucculent ftrings, as big as a goofe's quill, that run right down from the great root into the ground, to fupply it with nourifhment; for the great root lies upon the furface of the ground : From it arifes a ftalk, about two feet high, with feveral yellowifh-green long leaves, growing alternately from each fide of the ftalk. From amongft the ftalks and leaves arifes a ftalk with its bloffom, jointed; and at every joint it hath a membranous roundifh yellow leaf, covering the ftalk to the next joint, and fo all the way to the top, where the joints are fhorter and thicker, making the flower of a long fpherical head; the leaves of the flower refembling hops, between which is contained a little berry or feed, as in hops. The root, preferved or candied, is an excellent ftomachic, warming and comforting; boiled in wine, with a little cummin feed, it eafes the pain of the ftomach, and caufes fweat; outwardly applied, mixed with cocoa-nut oil, draws out poifons in wounds; and rubbed upon the ftomach, comforts it, and eafes pains from a cold caufe.

GLAND-FLAX, *or* NUIL.

This plant grows in great plenty in Peru, and is there called *nuil.* Its flowers are all white. The natives drink a tea of it, in ftoppage of urine proceeding from the gravel; it is alfo good to expel wind. It grows on mountains and dry places.

Another fort is the white *gavilu,* with a yellow lip,

E   grows

grows three feet high, in the same place with the last, and has the like virtues.

A third sort is *piquicken*, with a greenish flower, grows on the sides of the mountain, a yard high.

The fourth sort is *gavilu*, with a large yellow flower; the Indian women, newly laid-in, mix the juice of this plant with their broths, to cause their milk to return; which it does in plenty.

GOLDEN-ROD.

American golden-rod hath a strong thick striated green stalk, as high as a man, with rough dark-green leaves, four inches long, and sinuated about the edges; towards the top of the stalk are many branches and twigs, sustaining a great many naked yellow flowers, like those of St. John's wort or rag-wort. It is called *virga aurea major*. It is a most noble wound-herb, restringent, and healing all sores and ulcers in the mouth, or in any part of the body. It stops all sorts of fluxes, inwardly taken; and yet it provokes urine in abundance, forcing away that tartareous matter which breeds the stone.

GOOSEBERRY.

There are several of this kind in America, but not one of them to compare with the European gooseberry. Sir H. Sloane mentions eight sorts in his catalogue, most of which are without prickles, except that sort which they call Spanish gooseberry, which grows like the English bramble. I met with one growing in the mountains in Jamaica; its flower was so exactly in the shape of a rose, that I took it for one, but it had no sweet scent. Its fruit is black, cooling, and laxative.

GOOSE-

Goose-Foot, or Sowbane.

This herb is fo called for its killing (as it is faid) hogs, or making a fow caft her pigs, the leaf refembling the membranous part of a goofe or duck's foot. It grows very plenty in America, along the fides of highways, in yards, and in fome ftreets. It is a ftrong rank weed, of a very cold and moift quality, even to poifon, as fome affirm. Matthiolus faith, it works upwards and downwards: But Lycus Neapolitanus gave it againft the ftrangury, and inflammation of cantharides. It is better outwardly than inwardly; for it makes a good poultice or cataplaim, with hogs fat, againft fwellings and inflammations.

Goose-Grass.

There are two plants in Jamaica which refemble *cleavers*, or goofe-grafs. The fpecific quality of *cleavers*, or goofe-grafs, is to ftop fluxes and heal green wounds. The juice of the herb and feed, taken in wine, is good againft the biting of the fpider called *phalangium*.

Gourds.

Thefe grow into divers fhapes; as long, round, crooked, ftraight, fome exactly in the fhape of glafs bottles, and of all fizes, from an ounce to gallons. I had one prefented to me that held nine gallons, and very round. I carried to England, in the year 1717, two gourds exactly round like globes, both of a fize, containing fix gallons each. All thefe have a bitter pulp and feed.

The fweet gourd always grows long, as two or three feet, and as big as a man's thigh, which is full of fweet pulp, that makes a pleafant fort of fweat-meat or preferve. The feeds are one of the cold fpecies. The

diftilled

diftilled water is good in fevers. The pulp, applied to the eyes, abates their inflammation, and all other inflammations.

GRANADILLAS.

Thefe are tribed among the paffion-flowers, being the largeft of them all. Its fruit is as big as a fmall mufk-melon, and of the fame fhape and colour; the infide next the rind is reddifh, but the pulp is white, with many black kernels or feeds. It is of a pleafant tartifh fweet tafte, very cooling in fevers.

GRAPES.

Both white and red grapes grow very well in America. We have alfo a wild grape that grows fpontaneoufly in the woods, like bunches of Englifh elder, and of the fame bignefs and colour, but of a very pleafant vinous flavour, which are very beneficial to the hunters, to cool and refrefh them; and when there is no fruit upon its vines, cut but a piece of about a yard long, fuck one end of it, and it will afford a great deal of pleafant refrefhing water. There is alfo the American fea-grape, that grows along the fea-fide, which hath a very large leaf in fhape of a horfe's hoof, and its fruit as big as the common grape, and, when full ripe, of a bluifh black. Until they are thoroughly ripe there is no eating of them, they are fo rough and reftringent, curing fluxes; and whenever fo ripe, they have a flipticity and roughnefs upon the tongue, and binding. They grow by the fide of the fea, and oftentimes in the fea and falt water, like the mangrove, and therefore are called, by fome, mangrove-grapes.

GRASSES.

We are not without grafs in America; few countries

tries abound with more forts, and thofe green all the year. Their tribe is one of the largeft of any of the vegetable fpecies, including cyprefes, fea-grafes, &c.

There are fourteen forts of dog-grafs, nine of the land, and five of the water. The effence or decoction of the roots opens obftructions of the liver, fpleen, gall, reins, and bladder, provokes urine, and eafes the cholic. The diftilled water is given to children for worms, but you would do well to give with it a drachm of the powder of female fern, and half a drachm of wormfeed.

There are alfo,

The millet or panic grafes, of feveral forts.

Scotch grafs, whofe ear is like the millet's: This they feed horfes with.

Bur-grafs, which is hearty food for horfes.

The trembling or quaking grafes.

The *dactylon*, or finger-grafes, of feveral forts.

One fort called Dutch grafs.

There are about fifteen forts of the cyprefs and galangal kinds, which are counted fpecifics againft peftilential difeafes, and expel poifon: There is both of the long root and round root.

The crofs grafs hath the fame virtues.

There are other forts which grow wild, that bear a fort of oats, which are hearty food for horfes, and are commonly called wild oats. There is one of thefe oat-grafes which is purging, which in Chili is called *guilmo*. They make a tea of its roots, which they drink hot in a morning, and which purges them like fenna; it grows in marfhy and wet grounds.

There are alfo feveral forts of fea-grafes: One is called the manattee grafs, from the manattee, or fea-cow, that feeds and lives wholly upon this grafs.

E 3   GREEN

Green Withe.

This runs up ftraight on the fide of a tree, by its little clavicles coming out at every joint, without twifting itfelf round the body of the tree.   It hath a green fucculent round ftalk, without any leaves.   Perfons that have loft the ufe of their limbs, by the nervous cholic, take this root and roaft or broil it over the fi e, bruifing it, and applying it to the wrifts, which gives eafe, and ftrengthens the limbs.   If you put a piece of this root into any liquor you defign to ferment, it fets a-working immediately.

Ground-Ivy.

It is good for coughs and catarrhs, and will difcufs tumours, for which it is admirable.

Groundsel.

This differs in nothing from the Englifh fort but in its leaf, which is more like the daify.   The juice, drank from an ounce to two, mixed with a proper vehicle, works upwards and downwards, cleanfing the ftomach and bowels of all corrupt and cold humours.

Guavas.

Baked or ftewed, they eat like Englifh wardens, and are as red.   The bark of the tree is very reftringent, and very commonly given in fluxes of the bowels.

Guinea-Corn, or Panicum.

So called from its great quantity growing all along the coaft of Guinea ; it alfo grows as well in America. It is of the millet kind, of which there are feveral forts : Some have red grain, and fome very white.   It is excellent food for man or beaft ; for the ftalks and blades cattle feed upon, as men do upon the grain.   I have

feen

feen a fort that the grains ftick clofe to the ftalks, whofe head or fpike was above a foot long, tapering to the top, full of very fmall grains or feeds, fet fo clofe together that it makes a long pyramid.

Guinea-Hen Weed.

This plant hath a very rank fmell, and when cattle eat it, their milk and flefh have a difagreeable tafte. The root, put to aching teeth, eafes them.

Gum Animi.

This gum is a fort of *cancamum*: It is clear like refin, of a white or whitifh-yellow, fat and oily. The cleareft and moft tranfparent, and of a fweet fcent when burnt, is the beft.

Gum Cancamum.

This gum refembles feveral forts of gums or refins, of different colours, clotted or fticking together, or of four different colours clinging together, iffuing from a tree of a moderate height: But the difference of colour arifes merely from the different ripenefs or age of the gum; for that which newly fprings from the tree, when cut, fhall be of a different colour from that which hath come out of the tree for fome time, which colour is owing to the fun and air, it being all the fame gum; as it is very common to have *lignum vitæ* gum of different colours, although all from the fame tree. The tree of this gum hath leaves like that of the myrrh. Lemery fays, it ftrengthens the ftomach and bowels, kills worms, opens obftructions of the fpleen, &c.

Gum Caranna.

This gum flows from the trunk of a tree like a palm, which grows plentifully in New Spain: Carthagena is

the

the only place to have it. It is fo famous a cephalic, arthritic, and vulnerary, that it is ufually faid, what *tacamahac* cannot cure, *caranna* can. It is hard, refinous, clammy, but not very glutinous; foftifh and tough; of a dark olive colour, inclining to a green; of a fweet fmell, and fomething aromatic in tafle. It is commonly wrapt up in plantain-leaves.

Hare's Ears

Are accounted panaceas for all forts of wounds, inward or outward. The juice, effence, or feed, given in powder to a drachm in a glafs of wine, is faid to refift the poifon of the rattle-fnake; and a cataplafm of the herb, applied to the bitten part, attracts the venom.

Harillo

Has a flower like broom, and leaf very fmall, of ftrong fcent, glutinous, and full of balm, which heals all green wounds.

Hart's Tongues.

They are of the fern tribe, having all the fame virtues and fpecific qualities.

Hawk-Weed

Is fo called from hawks, as is faid, making ufe of the juice to clear the eye-fight of their young ones; but which fort they ufe, there being many, botanifts have not yet fatisfied us. Thofe in America have a fmall fibrous root, from which fprings one round ftalk three or four inches high, with little bunches with long narrow leaves, their edges hairy, and their under-fides fpotted with blackifh fpots; at the top of the branches ftand yellow flowers, like thofe of European hawkweeds. They are reckoned cooling, drying, and aftringent,

gent, and therefore stop fluxes. The juice, with honey and roche-alum, makes an excellent eye-water.

HEDGE-HYSSOP.

There is a yellow-flowered hedge-hyssop grows in Chili. The Indians eat this herb in their soups, to refresh them. It grows in moist places, near rivers, two feet high.

HELICHRYSUM, *or* GOLDEN CUDWEED, GOLDEN TUFTS, *or* LOCKS.

It hath a woolly stalk, with many long narrow leaves, green on the upper side, and hoary and woolly on the under side; the flowers grow on the tops of the stalks, in tufts, without any foot-stalk; the outward leaves, or *capsula*, are like silver scales, inclosing the flowers, of a pale-purple colour, with yellow thrums as in daisies; then follow many pappous seeds, as in others of the kind. The whole plant is drying and restringent, which makes it good against all sorts of fluxes and catarrhs. It is good in quinsies, and all ulcers.

HERCULES.

This sort of prickly wood is set thicker and fuller of protuberances and prickles, which are also much longer, than the other sorts, so that they look like Hercules's club, and it is therefore called Hercules. The wood is very yellow; its blossom is almost like the cassia fistula; after which comes a short flat pod, in shape and bigness of a man's thumb: It is first green, then red, and, when full ripe, very black, containing three or four flat seeds, like the Barbadoes flower-fence. The root of this tree, finely scraped, and applied like a poultice to the foulest ulcer, will cleanse and heal it; as hath been often experienced, and first discovered, by negroes.

Hoe-

Hog-Gum.

This gum and its ufes are well known in Jamaica. It is fo called becaufe hogs, when wounded by the hunters, run to the tree, lance the bark, and rub themfelves with the juice, which not only prevents flies coming to the wound, but alfo heals it. The juice, when it firft comes out, is of a yellowifh-white, and then turns more yellow, and afterwards black, hard, and brittle, like refin. I muft confefs I do not know the tree itfelf, but have made great ufe of its gum; and know by experience, that, inwardly taken, it is an excellent thing in the belly-ache or cholic: *Take the juice, when new and frefh gathered, two fpoonfuls; mix it with as much water, fweetened with fugar;* drink it, it will give eafe immediately, and, in four or five hours, give four or five ftools; it is alfo good to put in a clyfter. When it is old, it is more of a binding and ftrengthening quality. Made into pills, and given after purging, it ftops a gonorrhœa. *Take hog's lard, four ounces; the fame of hog-gum; bees-wax, two ounces; yellow refin, one ounce; round birthwort-root in powder, two ounces; mix, and make a balfam:* This is a univerfal balfam to cleanfe old ulcers; it heals them and all green wounds. A plaifter of the hog-gum alone eafes the gout, and ftrengthens the part.

Hog-Weed.

This plant is vulgarly fo called by the planters in Jamaica, becaufe they feed their hogs with it, who eat it very greedily. It is of the valerian kind. They are cooling and moiftening plants, full of juice, like purflanes, having much the fame virtues.

HOLLY-ROSE, *or* SAGE-ROSE,

Vulgarly fo called; in Latin, *ciftus*. Sir H. Sloane mentions a fort in Jamaica, which I met with growing very plentifully in fome of the pooreft ground. It hath a ftem as big as a man's finger, covered with a reddifh-brown bark, fmooth, rifing three or four feet high, with brancoes towards the top, putting forth hoary or woolly leaves, deeply cut or jagged on the edges like nettle, about three inches long and one broad, having a ftrong fcent like the common *ciftus;* between which come the flowers, ftanding in a pentaphyllous calyx, being very large, and of a yellow or orange colour, with five or fix leaves like the *ciftus,* or like the wild canker-rofe; after the flowers, comes a fmall fhort head, made up of three ftrong cartilaginous membranes, in which are feeds, pretty large. I have obferved, as I have travelled along the roads, that the flowers opened wide juft at eleven in the forenoon. *Ciftus* and dwarf *ciftus* are drying and binding; they have a bitternefs, and a little heating upon the tongue, which are reckoned fpecifics for all forts of fluxes.

HONEYSUCKLE, *or* UPRIGHT WOODBIND.

I never could meet with any that was exactly like thofe in England, either in flower or fcent. There are feveral forts: The firft hath a green round fmooth jointed ftalk, and at every joint it hath a leaf, whofe foot-ftalk encompaffes the ftalk at the bottom of the joint, like a round cup, which fometimes contains water; the leaves are five inches long and two broad, fmooth and thin like the leaves of gentian or fpiderwort, and have feveral fmall white flowers, upon long jointed ftalks, refembling other honeyfuckles; after which follow feveral large round black *acini*, cluftered

very

very clofe together, making one berry; in each of which lies one black feed, in a thick pulp, which dries away. It is a mountainous plant.

The other forts will be mentioned under the title of *Wild Sages.* The fruits, leaves, and flowers, are of one and the fame effect; which is faid to confume the fwelling of the fpleen, and to procure a woman's fpeedy delivery. The diftilled water of the leaves and flowers is good to clear the face of morphew, fun-burns, and freckles; a decoction heals ulcers. The oil of the flowers is good againft crampnefs, numbnefs, and palfy.

Horse-Tail.

I have feen the very fame fort in America, by river-fides, as grows in England. It is a fpecific in ftopping fluxes of blood, whether inward or outward; and heals ulcers and excoriations, if you dry it and powder it, ftrewing it upon the part; it is alfo good in coughs and catarrhs.

Hound's Tongue.

This grows in the moft barren parts of South-America, where it is called *ylo.* It grows about two feet high, with blue flowers. In fome parts of Peru, they have no other fort of fuel to clear their quick-filver from their filver, and to melt it down.

Indian Shot.

This is fo called from its feed being round, black, and fo hard, that, blown through a trunk or pith, it will kill fmall birds; they are drilled through, and ftrung to make beads and bracelets. It grows exactly like the Indian arrow-root, only the flower of this is of a moft beautiful fcarlet colour. The leaves are

cooling

cooling and cleanfing; applied to the hypochondres, with water-lily and aninga-oil, they abate the hardnefs of the fpleen. The juice of the root corrects the corrofive poifon of mercury fublimate; dropped into the ear, eafes pain; and, mixed with fugar, and applied to the navel as a cataplafm, cures a diabetes.

Indigo.

This plant is called *nil* and *anil*. It is a fmall plant, that grows about two feet high, hath a blue greenifh ftalk, whofe leaves are more blue, fmall, and roundifh, about the bignefs of fenna; the flowers are very fmall, fpiked, and of a purple or reddifh colour; fucceeded by a fmall crooked pod, about an inch long.

The way of making indigo is fo difficult, that many planters never obtain it: I had a Papaw negro that would make indigo with any man in Jamaica; and, when they mifcarried in making it, would fend far and near to know of him the reafon, and to remedy it: I muft confefs I never pretended to direct him. The whole dependance is in due fteeping the weed, and beating its liquor. Now, knowing few in Europe know how indigo is made, no more than they do fugar, I will juft give a fpecimen of it. The feed is fown in rows by a line, and, if they have good feafonable weather, that is moderate fhowers of rain, the weed will be fit to cut in fix weeks time, which is done with a crooked knife, in the fhape of a fickle, but not jagged, and are called indigo-hooks. Then they have three vats or cifterns, into which they put the weed, and prefs it down with their feet as clofe as they can; and, when full, they lay large fticks over it, which are preffed down with beams that go acrofs the cifterns, faftened in a poft in the ground, four or five feet deep; all which is to keep the weed from rifing up when they put water

ter

ter to it, which they do as much as it will imbibe, and over-top it five or fix inches; which, in twenty-four hours, will grow fo hot that you cannot put your hand into it, and it will boil and bubble like a pot boiling over the fire, and the water be tinged of a blackifh-blue colour. When the weed is fteeped fo long that it begins to rot, then they let go the water from it into another lower adjoining ciftern, where it ftands about twenty-four hours; and then they beat or churn it very well for three or four hours, until its grain appears, and feparates from the water. The way of beating is by a pole, with boards made tapering at the end of the poles, bored full of holes, which they beat or churn the liquor with; and when it is near finifh-ing, they take fome of the liquor, and put it into a porringer, and let it ftand, to fee how the grains, or fine muddy particles, precipitate to the bottom; which if it doth well, and the top looks clear, then they leave off, and let it ftand twenty-four hours longer, for the mud to feparate from the water. They then have two or three tap-holes, to let out the water into another fmall fquare hole, which runs out as long as it runs clear. Then they lade out all that water, and let out the mud into the fquare hole; which they put into bags made tapering, or like Hippocrates's fleeve, of coarfe oznaburghs, and let the water drain from the mud as long as it will drop. Then they empty the bags into a fquare frame, ftir it well together, and dry it in the fun. If they make it into flat cakes, they have boxes on purpofe; if into lumps, which they call fig indigo, they put fpoonfuls, or lumps, upon a cloth ftretched out, and dry it in the fun.

Befides the common indigo, there is another fort called wild indigo, whofe leaves are much fmaller than the former, but is more hard and woody, growing fome-

times

times eight or ten feet high; whofe ftalks are of the
bignefs and colour of Englifh broom, but the flower
and feed are exactly like the former.

IPECACUANHA.

There are four forts of thefe admirable purging
plants.

1. The black fort, which hath a fmall, crooked,
knotty, and wrinkled root, almoft like *afarabacca*, but
not quite fo big; from which arifes a fmall ftalk, of
about half a foot long, partly creeping, and partly rifing up, adorned with a very few leaves, which are like
thofe of the pellitory of the wall. In the middle of
them grow five-leaved white flowers, upon a little foot-
ftalk, and a capfula almoft like a rofe; after which
come reddifh-brown berries, of the fize of a fmall
cherry, and black when ripe; within is a white juicy
pulp, inclofing two yellowifh feeds, hard, and in fhape
of a lentil. The root, when frefh gathered, is of a
dark-brown colour, of an unpleafant ftrong fmell, a
hot and fharp bitterifh tafte, and, when well dried, will
keep for many years. The firft of it that was brought
into France was in the year 1672, by Monfieur le
Gras, a phyfician, who had made three voyages to
America, but concealed the name of it, and called it
the Indian root. After him, Abbot Bour de Lot made
ufe of it; but Helvetius (however he came by it, or
the knowledge of its virtues) was the only perfon that
was famed for the ufe of it in all forts of fluxes of the
bowels, which gained him great reputation for the curing that diftemper: Upon which, after the king of
France was truly fatisfied of the great cures he performed, he purchafed the knowledge of his medicine,
which proved to be this root. The king fatisfied Helvetius for his difcovery, appointed him phyfician to
one

one of his hofpitals, and made the root known to all
his fubjects in France; where it was fold for many
years at 3 *l. per* pound.

2. The other root is like this, but whiter and
weaker.

The ufe of both thefe roots is to cure dyfenteries
and diarrhœas, that is, bloody and other fluxes of the
bowels, by removing the tenacious morbific matter
from the part affected; expelling it by vomit, and
fometimes by ftool; after which it aftringes, binds, and
ftrengthens the tone and faculty of the bowels, refto-
ring the perfon to his former health.

3. The *caapia*, whofe root is thick, foft, and ver-
rucous, like the other fort, and full of tender fibres or
filaments; from which root fpring three or four round
ftalks, having but one fingle leaf, of a bright-green on
the upper fide, and on the under a little whitifh. It
bears a flower like a daify; round as a navel, on a
fingle ftalk, with many fmall ftadles, which form themfelves into a berry, containing a feed lefs than muftard-
feed.   The virtue and efficacy of this root is the fame
with the aforefaid roots; but it is accounted more an-
tidotal, expelling the poifon of *aconite* or wolfsbane,
and other like forts of poifons.

4. The *caatagá*. This excellent plant fcarce rifes
the height of a hand, with one tender four-fquare
ftalk, which is always green, partly lying on the ground,
putting out at the joints fmall roots, which run again
into the earth; at which joints there come out two
fmall leaves, ftanding oppofite to each other, in form
and bignefs of money-wort, but rougher, and jagged
on the edges like vervain, fpeedwell, or germander, of
a pale-green, and at every pair of leaves are very fmall
white hooded flowers; after which come the feed-veffels, in bignefs and form of oats, which opening of
themfelves

themfelves fhed a very fmall round yellowifh feed, lefs than the fmalleft poppy-feed. The whole plant hath little or no fmell, but is of a bitter tafte. It grows generally in meadows and moift favannas.

The roots of this plant are very fmall and fibrous, and of a fub-bitter tafte. The more experienced inhabitants of Brazil efteem it as one of their moft noble purging plants. The juice or a decoction of the plant, or the root itfelf in powder, given in a fmall quantity, fo ftrongly moves the menfes, that it is not fafe for women with-child to take it, becaufe it purges by ftool at the fame time. The dofes of all thefe forts of roots are from a fcruple to two, drinking with it green tea, or thin poffet drink. You may alfo gently infufe the root in warm water, which pour off, and that will gently purge; and the remaining root, dried and pulverized, is more fit for weak perfons for all the aforefaid purpofes.

IRON-WORT.

We have a fort of this plant growing in Jamaica. It hath a four-fquare ftalk, rifing to about three feet high, from a white fibrous root. From each fide of the ftalk come out two leaves, oppofite to one another, exactly like iron-wort, and of the bignefs of fig-wort. Towards the top come out, all the way from the foot of the leaves, fpherical heads, as big and like wild hops (and therefore fo called by fome), made up of a great many white flowers, ftanding clofe and round together, upon an inch-long foot-ftalk, like the meadow purple trefoil; after which follow many fmall black fhining feeds, which make the whole head fhew black. This plant hath a fpecific quality to heal all wounds, and ftop all fluxes of blood and other humours. A decoction of this plant, with honey, makes an excellent mouth-water, and for fore throats.

F JABORAND.

Jaborand.

In America are divers forts of this plant. The firſt fort hath a yellowiſh crooked root, full of fibres, and in fmell and taſte is like the pellitory of Spain; from which root ariſes an aſh or grey coloured tender ſtalk, running upright for a little way, and then dividing it-felf into branches, putting out upon a foot-ſtalk three leaves, ſharp-pointed like a ſpear, with many veins of a pale or whitiſh green, ſeeming rough and hairy, but foft to the touch. It hath a tetrapetalous or four-leaved flower; after which follows the ſeed, in a double capfula, like hemp, compreſſed, and in ſhape like a heart. This plant is a great antidote againſt poiſon; and Piſo affirmeth, that a captain in Brafil, who was poiſoned by eating venomous muſhrooms, was imme-diately cured by a native Indian, in the preſence of the Prince of Naſſau, only by taking the juice of this plant.

The fecond fort is like the firſt, but only the leaves are much larger, and of a deeper green colour, and the ſtalk more knotty or reaved at an equal diſtance. The virtues and uſe are much the fame.

The third fort is like the *betys*, which fome call Spaniſh elder. The fourth fort Sir H. Sloane calls *piper longum*, &c. Theſe two forts are of the fame virtues as the firſt, but not ſo powerful.

Jalap.

Theſe roots are called *mechoacan:* There are two forts, white and black; the black is moſtly uſed, and is called jalap. It differs little or nothing from the four o'clock flower; but it is certain that this fort hath more refin in it than any other *mechoacans*, which is the purging quality, and therefore of more uſe. Chooſe that which is heavy, cloſe, and ſhining when broke, which is the refin part.

White

White jalap is much larger and whiter than the black or common fort, and is a convolvulous plant, climbing upon trees. It hath a milky, knotty, reddifh, multangular ftalk, having here and there folitary leaves, which are tender and very green, in fhape of an heart, fometimes with earlets, and fome of the leaves without. The flowers are monopetalous and tubular, with four indents, and of an incarnate or pale rofe colour with-out-fide, and within, towards the bottom of the flower, of a purple colour, ftanding in a calyx; and after the flowers come the feeds, as big as peafe, but a little compreffed as if triangular, contained in a ciftus which ftands out like a navel. The root, on the outfide, is brown, a little rough, of an oblong fhape, and large. Thefe roots are cut longways, whereas the black jalap is cut croflways; thefe purging, but not fo ftrong as the black fort, therefore of lefs ufe, but of the fame virtues. A gentleman affirmed to me, that by only holding this root a little while in the hand, it took away the cramp, and never failed; and people ufed to come to him for the cramp-root, not knowing what it was, or by any other name.

JESSAMIN.

There is a great variety and plenty in Jamaica. The true or wild jeffamin is a very large tree, growing wild in woods. It hath long large thick leaves, in fhape of a large laurel, with a milky juice, and hath white odoriferous flowers, perfuming the very woods they grow in: But the Arabian jeffamin, which hath a fine white flower, like to orange or lemon flowers, exceedeth all the reft in fweetnefs. Jeffamins, diftilled, make a beautiful wafh, and perfume at the fame time. The inhabitants make a wonderful fweet oil of jeffamins, and, with the mixture of fome other odori-

F 2   ferous

ferous powder and balfams, make it as ftiff as bees wax, forming them into feveral fhapes and colours, which the Spanifh ladies put into filver or gold boxes, and wear them about them. They will keep good many years, and are of great value.

KETMIA.

Thefe have moft of them a mallow leaf, and are therefore tribed amongft the tree-mallows, or *alceas*. One fort hath a mallow leaf, and is of an acid tafte, like forrel.

See Sorrel.

LACAYOTA

Is a fort of lemon-balm, which lafts green all the year. It makes fine arbours, running up to the tops of houfes; and is an excellent preferve. It is much like the water-lemon.

LAGETTO TREE

Hath a laurel-like leaf. The inward bark may be drawn, only by pulling it with the fingers, into the fineft lace that can be wrought with needle and thread, of what breadth and length you pleafe; it will alfo bear wafhing with foap, or currato, as white as other lace. The negroes and Indians make fine white ropes of it; and I am perfuaded, that fine cloth may be made of it, which might turn to great profit and ufe, if people would take pains to improve what nature offers.

LANCE-WOOD.

So called from its ftraightnefs and toughnefs. Negroes make lances of it, and it ferves for rods. The pigeons feed upon the berries, which make them very fat.

LAURELS.

Laurels.

There is great variety of the laurel kind, or of those trees that have a laurel leaf; but as they are moſtly known by other names, I ſhall ſpeak of them as they come in their turn.

Lavender.

We have two or three ſorts in America, ſome odoriferous, others without ſcent. The ſeed and leaves of the plant are excellent in fits of the mother.

Lemons.

Beſides the common European lemon, which we have in great plenty and as good as any in the world, we have a water-lemon, which hath a fine large paſſion-flower, and is therefore tribed among them. Its fruit is a moſt pleaſant ſweet with four, and mighty cooling and refreſhing in fevers. It is a climber, and makes ſuch thick arbours that you can hardly ſee through them.

Lentils

Are a ſort of vetch, or ſmall pulſe, of which there are many ſorts that grow wild in America.

Licti, *or* Luisi Plant.

In Chili, there is a very common tree called *licti*, the ſhade whereof cauſes the bodies to ſwell of thoſe that ſleep under it; but more eſpecially the face, ſo that they cannot ſee out of their eyes. The ſame doth a ſhrub that grows in Providence, and in Bermudas, called the poiſon-ſhrub; for if you do but go to the leeward of it, the wind will drive its malignant poiſon upon you, ſo that a great itching and ſwelling all over

your

your face and hands will enfue immediately. Its berries alfo are a ftrong poifon : To cure the diftemper, they take an herb called *pellbogui*, which is a fort of ground ivy, that bears a berry as big as a winter-cherry, which they pound with falt, and rub the part affected ; by which means the fwelling goes off in two or three days, fo that no ill remains.

Lignum Aloes.

I met with a tree in Jamaica that had a very black heart, and a fine fcent, much refembling lignum aloes, being very bitter : A carpenter who firft fhewed me this tree, called it fweet iron-wood. A negro that I employed to get fome of it, when he brought it me, faid the fame fort grew with them in Africa, where they called it *columba*. The fineft is the black kind, clofe and heavy. It hath leaves fomething like the olive ; after which grows a little round fruit, like the Englifh cherry. There are three colours of it imported into Europe : The firft hath a very thin bark, and under that is a very black heart, clofe and folid like ebony ; the fecond fort is a light veiny wood, and of a tanned colour ; the third fort is all folid heart, and is called the precious wood of Tambaok. Choofe that which is fhining, as green as a leek without, and of a light-yellow within, bitter in tafte (from whence it hath its name), and will burn like wax, yielding a fweet fmell. It kills worms, and is cephalic, narcotic, ftomachic, cardiac, and alexipharmic.

We have a wood called iron-wood (for its durablenefs, hardnefs, and lafting), having a very black heart.

Lignum Rhodium, *or* Rose-Wood.

The negroes corruptly call it *lignum rorum*; by fome it is called candle-wood, becaufe it burns like a candle,

and

and fmells very fweet, being full of oil. There is another fort of black candle-wood, which I take to be the lignum aloes. Thefe are of the laurel-leaf kind.

LIGNUM VITÆ.

Thefe trees grow in all or moft of the woods in America, and are known by all its inhabitants. Its flowers are five-leaved, and of a delicate blue colour, of which may be made a purging fyrup, like fyrup of violets. The fruit is very purging, and for ufe excels the bark: Were it known in Europe as well as here, they would never ufe the bark or wood, but its fruit. I have cured venereal difeafes and yaws with this fruit, without falivation. The gum is a moft admirable medicine; and the ufe of it is fo well known in Europe, by the name of *guaiacum*, that I need not fay more of it. This tree is one of the ever-greens.

LILIES.

Of thefe there are many forts in America: They all have the fame virtues as the European lily.
See Water-Lilies.

LINE, *or* LINDEN-TREE.

There are in Jamaica two forts of thefe trees. A decoction of the leaves cleanfes and heals fore mouths and cankers, and takes away fwellings in the legs.

LIMES

Are a diminutive lemon. They are fo common that the planters fence their plantations and paftures in with them; the fruit is generally ufed, in the room of lemons, to make punch with. The negroes and Indians ufe the root in venereal cafes, and the ftalk to clean their teeth with.

LIQUID

Liquid Amber.

Is a natural balfam that flows out, by incifion, from the bark of the trunk of a large tree, whofe leaves are like the ivy. Its bark is thick, of an afh colour, and very odoriferous, fmelling fomething like amber-gris, from whence it hath its name. The Indians call it *ococol*, or *ocofols*. They are plenty in New Spain. The Spaniards call this gum *matricalis*, for its great virtues in womens diftempers, as hardnefs of the womb, opening obftructions; it prevents hyfteric fits, and cures the *fiftula in ano*, and all other wounds. I have known defperate ulcers of the throat cured by it, and quinfies and fore mouths. It is good in *fciatica* rheumatifms, weaknefs of nerves, and contracted finews.

Liquorice.

I have feen the European liquorice grow very well in gardens; but we have two forts of plants that have a liquorice-tafte: The one is a vine, whofe leaves have the true tafte of liquorice, and is therefore called wild liquorice; it bears a red fruit, in fhort pods like peafe; it winds itfelf round any fhrub it comes near, rifing to the height of the tree; the ftalks are about the bignefs of a goofe-quill, fet with winged leaves, of equal number on a fide, oppofite to one another; it hath a fpiked clofe papylionaceous flower, of a pale purple, and is pea fafhion; after which follow fhort greenifh pods, but black when dry, which contain three or four fcarlet peafe, with a black fpot on that part it fticks to. Thefe grow in both the Indies. In the Eaft-Indies, they make necklaces of the fruit or peafe, which they fay prevent the children that wear them from the fever, make them breed their teeth eafy, and prevent cramps and convulfions. They are of a more beautiful red than

red

red coral; and, if for nothing elſe, they make beautiful necklaces. I knew a gentleman in Jamaica that
made a tea of the leaves, and drank of it many years,
which he ſaid kept him in good health. I have often
ordered a ptiſan of the leaves with good ſuccefs in
cholics. The root of this plant, although it hath not
the taſte of liquorice, yet it hath the colour, both outſide and inſide, of Engliſh liquorice-root. I have
obſerved ſheep to feed greedily upon its leaves.

The other wild liquorice is a ſhort upright ſhrub,
which ſome call ſweet-weed. The whole plant taſtes
like liquorice. Its ſtalks are hexangular, branching
out every way like a little tree, about a foot and an
half or two feet high, beſet very thick with leaves, three
at a place, without foot-ſtalks, and about half an inch
broad and three quarters long, ſerrated about the
edges, and of a graſs-green colour. *Ex alis foliorum*
come the flowers, on a quarter of an inch ſtalk, which
are whitiſh-blue, and tetrapetalous, with many ſtamina
ſtanding round; then follows a little round head, or
ſeed-veſſel, not much bigger than great pins heads,
containing ſuch ſmall brown ſeeds as can hardly be diſ-
cerned. Three ſpoonfuls of the expreſſed juice of this
plant, given morning and night for three or four days,
is an infallible remedy for a cough.

LIUTO

Is the name that the South-American Indians give
to a flower like the flower *de lys*, although there be of
them ſeveral colours; and of the ſix leaves that com-
poſe it, there are always two crowned. Of the root
of this flower, dried in an oven, they make a very
white meal, and paſte for confectionary.

LIVER-

Liver-Wort.

This plant I found growing in great plenty about a mine at St. Faith's in Jamaica, on the shady banks of the river-side, and also about the hot springs to windward. Liver-worts are so called from their great virtue in curing diseases of the liver, and consequently are good in the jaundice. They gently purge choler; bruised, and boiled in beer, and drank plentifully of, they help in a gonorrhœa and female weakness; outwardly applied, are said to cure malignant scabs, tetters, and ring-worms; and to cleanse and heal old ulcers.

Locus-Tree.

It is also called *lotus* tree. We have three or four sorts of them.

1. This tree hath a very beautiful reddish flower; its fruit is about the bigness of the American clammy cherry, of a yellowish colour, and very pleasant to eat, which men and birds covet; but they have much stone in them, which is the seed. The bark of it cures intermitting fevers as well as the jesuits bark, as I have often experienced, and that in the same proportion or quantity; and the bark taken from the limbs and branches is of the same colour, in all respects, as the Peruvian quill-bark, which is reckoned the best. I knew a practitioner of physic in Jamaica, who used no other for many years in fevers, but kept it as a secret from what tree he had it, most people supposing it was the bully-tree bark, but he affirmed to me that it was not. I at last got out of a negro, that used to gather it for him, what tree it was, which I found to be this sort of locus (there being two or three other sorts); upon which I used it in intermitting fevers, in the room of jesuits bark, with the like success.

2. The

2. The flowers of this tree are more yellow, and its fruit much fmaller, but of the fame nature.

3. The firft I ever faw of thefe trees was about twenty-feven years paft, at one James Pinnock's, at Liguanea in Jamaica, who told me it was a Barbadoes locus-tree: It was a fine large fpreading tree, in big-nefs and fhape of the Englifh beech-tree. The fruit is broad and thick, with a hard fhell, and about fix inches long, of a cinnamon colour; wherein were three or four round flat blackifh beans or ftones, bigger than thofe of the tamarind, inclofed in a whitifh fubftance of fine filaments, as fweet as fugar or honey. When frefh gathered, it is faid to purge; which quality it lo-feth as it grows old. The juice or decoction of the leaves expels wind, and eafes the cholic pain, by giving a ftool or two. The inward bark deftroys worms in young or old.

Logwood.

It is often called Campeche-wood, from the great quantity growing in the Bay of Campeche, where the Englifh cut it, and fend it to Jamaica; but not with-out great rifk and hazard of their lives, being in the dominions of the Spaniards, who often cut them off. In the year 1715, I had an Indian flave, that I fent down to the Bay of Campeche to cut logwood, whom I ordered to fend me up fome of the feed of it, which he did; and I ordered it to be planted in Jamaica, where it takes to growing admirably well, even in the worft of the lands; fo that there are now feed-bearing trees enough to ftock the whole ifland; and, in a little time, the Englifh need not run thofe rifks as formerly in cutting of this wood, which they ufed to do ftanding up to the knees in water, with the mofquitoes lancing and tearing their flefh, by which many thoufands died,

befides

befides every day running the danger of being cut off
by their enemies for robbing.　Its leaves are much of
the fhape and bignefs of *lignum vitæ*; its feed is in a
thin membranaceous cafe, hanging in bunches like the
Englifh afhen-trees.　A decoction of the wood ftops
bloody and other fluxes.　This is one of the dying
woods.　They now make fences of them in Jamaica,
which are fo thick and prickly that nothing can pafs
through them, and, being an ever-green, you can
hardly fee through them.

LOOSE-STRIFE.

The American loofe-ftrifes much refemble thofe in
England, and have the fame virtues.　They are excellent wound-herbs; ftop bleeding, inward or outward; cure fore throats, fore eyes, and venereal ulcers.
The juice or effence ftops fpitting of blood, and bloody
fluxes.　A cataplafm or ointment, made of this plant,
is an excellent balfam.　The diftilled water is a cof-
metic.　The whole plant, made into fmoak, drives
away mofquitoes, &c.

LOVE-APPLES,

So called by the Spaniards, who ufe them in their
fauces and gravies; becaufe the juice, as they fay, is
as good as any gravy, and fo by its richnefs warms
the blood.　The fruit of the wild fort is no bigger than
a cherry; but thofe that grow in gardens are as big as
a fmall apple, very round and red, and therefore called *pomum amoris*; fome call them *tomatoes*.　It hath
a fmall fharp-pointed jagged leaf, growing very thick
upon its ftalk and branches; its fruit is round and red,
or of an orange colour.　I have eat five or fix raw at
a time: They are full of a pulpy juice, and of fmall
feeds, which you fwallow with the pulp, and have
　　　　　　　　　　　　　　　　　　fomething

ſomething of a gravy taſte. Its juice is cooling, and very proper for defluxions of hot humours in the eyes, which may occaſion a *glaucoma*, if not prevented; it is alſo good in the St. Anthony's fire, and all inflammations; the fruit, boiled in oil, is good for the itch; and a cataplaſm of them is very proper for burns.

LUCIMO.

In the province of *La Sarena*, in Chili, and Peru, there is a tree which is called *lucimo*. The leaf of it ſomewhat reſembles that of the orange-tree, or *floripondio*; the fruit alſo very like a pear; when ripe, the rind of it is a little yellowiſh, and the fleſh or pulp very yellow, with a little bitterneſs; in the midſt is a very large rough kernel or ſtone, bigger than the avocado pear. Theſe are called in Jamaica *mammees*.

MACAW-TREE,

So called from a large bird that feeds upon the fruit of this tree; which is of the palm kind. There are two ſorts of them, but they differ in nothing but the fruit; there is one bigger than the other. This tree is full of ſharp prickles from its bottom to the top, and all the ſtalks of the branches, which are exactly like the common palm. It hath a black flat round nut, in ſhape and bigneſs of what is called here the horſe-eye bean, covered over when ripe with a yellow pulp, like the common ſmall palm, which the macaw greedily ſwallows. The outſide part of the body of the tree is exceſſive hard; of which the Indians make their bows, and ſeveral other uſeful things; but the inſide is full of a ſoft pithy ſubſtance, like the cabbage-tree.

MAD APPLES.

Theſe are tribed among the *ſolanums*, or nightſhades;

ſhades; they are vulgarly called *valanghanna*, in Jamaica. The only reaſon, that I can find, why they are called mad apples is, becauſe they bear ſome reſemblance to mandrakes: Some have fancied they were the male mandrake, and, imagining them to be poiſonous, did for that reaſon call them mad apples: But I know by experience to the contrary, having eaten many of them, both boiled and fiied; but the beſt way is to parboil them, taking off their outer ſkin, which is a little bitteriſh, and then fry them in oil or butter. I planted, above twenty years ago, half an acre of ground with them, on which my ſlaves fed, and were well pleaſed with the food. They eat ſomething like a ſquaſh, but better than any of the pompion kind; and are ſo well known in America, as to need no particular deſcription. Angola negroes call them *tongu,* and the Congo negroes *macumba.*

Maguey,

Of which they make a fine thread called *pita,* and we call it ſilk-graſs. It is certainly one of Piſo's *caraguatas;* who ſaith, from *maguey* they get honey, vinegar, and drink. Now it is certain, that excellent drink may be made from the pine; and I believe the juice, being ſo ſweet, may be boiled up to a ſweet extract like honey, and alſo its ſweet juice, after fermenting, will turn to good vinegar: But of the penguin the natural juice is ſharper than any juice of crabs, lemons, limes, or the ſharpeſt vinegar; and the fruit ſome will eat of until they fetch the ſkin off the tongue and make it bleed. Now he ſaith, the ſtalks and leaves are good to eat, but none of them can be eaten; they do indeed make a fine thread. The wood, he ſaith, ſerves to cover houſes; but neither of theſe are fit, for they have no wood: Its prickles or thorns for needles;

neither

neither of thefe hath fuch long fharp prickles, but very fhort crooked ones : And the Indians ufe the fruit in-ftead of foap. By this it fhould be the *caraguata* that we call currato; for the leaf of that (not the fruit) many people ufe as foap for their linen; and almoft every houfekeeper ufes it to fcour their bowls, difhes, plates, and floors. It hath a very large ftem, as big as a man's leg, that they may cover their houfes with; but it hath not prickies fit for needles : So that I am at a lofs which of thefe three forts to affign the Chili *ma-guey* to. It is certain, nothing of the currato can be eaten. Some Mexicans call the currato *maguey*, and the penguin *maguei prunorum;* fo that it feems moft probable to be one of thefe, but chiefly this of curra-toes, for fcouring like foap.

MAHOTS.

The firft I ever faw of thefe trees was above twenty years paft. Walking by the Rio Cobre, near St. Jago de la Vega, in Jamaica, I obferved a very beautiful large fhady tree, full of green leaves, large and round, ftuck full of fine red flowers; and, upon a ftrict exa-mination, I found it to be one of the tree-mallows. Its flower has a little fweetifh fcent, but in exact fhape and colour of the red lily.

The fea-mahot, with a yellow flower : The bark of thefe makes fine white ftrong ropes.

The bark of thefe trees is often called *maho*, from the corruption of *mahau* and *mahot*, *&c.*

MAIDEN-HAIRS.

There are many more forts of maiden-hairs in America than in Europe, and fome of them much larger; while others, both golden and black maiden-hairs, are exactly like thofe of England. They are fpecifics

cifics againſt all obſtructions of the lungs, liver, ſpleen, &c. and heal and dry ulcers.

MAJOE, *or* MACARY BITTER.

This admirable plant hath its name from Majoe, an old negro woman ſo called, who, with a ſimple decoction, did wonderful cures in the moſt ſtubborn diſeaſes, as the yaws, and in venereal caſes, when the perſon has been given over as incurable by ſkilful phyſicians, becauſe their Herculean medicines failed them; *viz* preparations of mercury and antimony. It is alſo called Macary bitter, from its growing in great plenty in the bay of Macary, and being a very bitter plant. I met with ſome of it growing in a ſkirt of a wood near St. Jago de la Vega, in Jamaica: It was but a ſmall tree that I ſaw, with winged leaves much like the Engliſh aſh; the flower I never ſaw; but the fruit is in cluſters, in ſhape and bigneſs of the Canary grape, firſt green, then of a bright ſcarlet, and when full ripe as black as a damaſcene plumb: It hath a yellowiſh pulp, with a ſub-bitter taſte; then a large ſtone, with a kernel or ſeed in it, all very bitter: This plant was firſt ſhewn to me by a planter, who had done many excellent cures amongſt his negro ſlaves, in old inveterate ſtubborn ulcers, and that by only boiling the bark and leaves, or flowers and fruit if they happen to be on the tree when wanted to make uſe of, giving them plentifully to drink, and waſhing the ſores with ſome of the decoction; then laying over them a leaf of the jack in the buſh, until their ſores were healed.

MALLOWS,

Of which there are many ſorts in America, are divided into three diſtinct claſſes.

1. The

1. The common mallows, whofe feeds ftick clofe to their outward membrane.

2. Are properly *abutilons*, whofe membranes are fomething laxer.

3. Are *alceas*, and are thofe whofe membranes or follicles are not difpofed as in the others.

But all of them are mucilaginous. Some of them I fhall fpeak of under other names; but as for the common mallow and marfh-mallows, their virtues are already fufficiently fet forth in every herb-writer.

The moft common mallow in America is a large hairy rough mallow, with a yellow flower. Of this I have often made an excellent mucilage, in order for *unguentum dialthea*, although we have the fame *althea* as in Europe.

There is alfo another mallow: Its flowers are yellow, with a purple fringed bottom, and cordated petals. A tea of this is wonderful in diforders of the ftomach. The Indians make a poultice of it, which they apply to ripen fwellings, and is counted an univerfal remedy. It grows in moift grounds, and by river-fides.

American mallows with an elm-tree leaf, and flowers ftanding in knots at the angles of the leaves with the ftalk.

American mallows with vine-leaves, and roundifh prickly fruit.

American mallows with the leaf and outward form of ground-ivy, and hufks or cells double-forked.

American mallows with an ivy leaf, and with a fcarlet red flower.

Downy American mallows, with the leaf of the mufk-melon.

Mallows and marfh-mallows have much the fame virtues; all allow them to be powerful emollients, and

G

to foften violent pain by their flimy juice or mucilage, not only blunting the points of the corrofive falts, but relaxing and foftening the fibres which undergo too great a tenfion, it reftores them to their ufual fpring and tone, and confequently allays the pain. The herb, root, and feed, allay inflammations, promote expectoration, and expel urine, flone, and gravel. A drachm of the root in powder, given every morning in milk, is an excellent remedy in a gonorrhœa.

Of the American *alceas*, the bark of one fort is as good as any European hemp, and this the Indians and negroes make ufe of. Alfo the long okra, the fhort round okra, the mufk-mallow, and the vine-forrel. *Alceas* have the fame virtues, although not in equal degrees, with mallows. The moft hairy fhrub vervain mallows of America. The hollyhocks are *alceas*, or large tree-mallows.

See Mahots, &c.

Mammee-Sapota.

This is a very beautiful tree, full of fine branches and long green leaves, but feldom grows above fifteen or twenty feet high. Its fruit is almoft as big, and in fhape of, a man's heart, only a little longer, and fharper at the lower end; the outfide is of a brown or ruffet colour, and very rough; the infide is a darkifh-red foft pulp, and lufcious eating, like a mamulet; in which are contained two, fometimes three, long cones or ftones, thick in the middle, and fharp at both ends, one fide rough, and would make good nutmeg-graters, and the other fide fmooth, black, and fhining as poffible. It is faid, thofe that plant the ftone or feed of thefe trees never live long enough to eat of the fruit of them, being forty or fifty years, as they fay, before they bear: I have feen one, that a perfon told

me

me he planted above twenty years ago, and there was no sign of its bearing then.

MAMMEE-TREE.

These are very large spreading trees. When cut, there comes out a yellowish gum, like *tacamahac*, which, applied to any part that hath chigoes, will draw them out whole, bag and all, sticking close as bird-lime. It hath splendid smooth leaves, and a large fruit, as big as an English custard-apple, of a buff-colour without-side, and yellow within; having one or two large stones within the fruit, very rough, and sticking to the pulp, some of which are very bitter, some very juicy and delicious, others hard, and of the taste of a raw carrot.

MANCHIONEEL.

There are three forts of them: One whose fruit is round, and in bigness like an English genetin, and which, when ripe, smells like them, and is very tempting to eat; of which some have to their cost, it being a fort of poison, but its milky juice is worse.

The second fort hath fruit in the shape of an heart, or a little pointed at the extreme end, like the great Seville or Spanish olive, and of the same bigness.

The third fort hath a very small round fruit, of the bigness of a small cherry. Its juice is corrosive like the other fort, but turns black immediately; whereas the others have a very white milky juice, which the sun hardens to a fine hard yellow resinous gum, not inferior in virtues to the guaiacum. That which hath the black corrosive juice is called by some hog-doctor, or the hog's doctor; for when the hogs are wounded, they run to one of these trees, and lance it, then rub the wounded part with its juice, and after that no fly or vermin will come near the fore. It is certain the fruit of thefe trees are poifon, infomuch that the land-crabs that eat of them, although they do not poifon the crab, yet thofe that eat of thofe crabs fhall be taken very fick; fome have died fuddenly after. Some of thefe trees grow by the fea and river fides; and it hath been obferved, that fifhes will eat of their fruit as they drop into the water, which will make their teeth turn yellow, and become poifonous. I had a negro man that wilfully poifoned himfelf with them, and a little before he died he confefled it, and would fain have lived: I obferved, he complained of a great heat and burning in his ftomach, but could not vomit; his tongue fwelled, and was burning hot, as he called it; he was continually calling for water; his eyes red and flaring, and he foon expired. It is faid, the Indians put the juice of this tree, which is more corrofive than the fruit, into the nicks or notches of their arrows, in order to poifon the wound the arrows make, that it may not be healed or cured. I have experienced, that if you lay thefe apples in a prefs where cockroches have got into, they will foon forfake it. Although the juice of this tree is fo venomous and fharp as to put out the eye immediately, when it hath happened to fly into it by cutting the tree (for which reafon they make fires round them, and fcorch them very well before they cut them), yet this venomous milky juice, in time, will turn to a fine refinous gum; which I have given inwardly, many times, as we do gum guaiacum, for the fame purpofes, and with the fame effect. Indeed, at firft, I ufed it for gum guaiacum, and it was fome time before I could find out the deceit of the negroes, who fold me one for the other, they were fo alike: But, after I knew the difference, and found no ill ef-fect, but the fame as if it had been gum guaiacum, I

then

then continued the ufe of it, generally diffolving it in a rectified fpirit of wine, making a tincture; and I defy the nicest perfon to know it from tincture of gum guaiacum; befides, its virtues are the fame. I have found it by experience to be a fpecific for the dropfy, carrying off all the watery humours by ftool and urine; only it muft be obferved, after the water is evacuated by this gum, to give a decoction of contrayerva and fteel, to ftrengthen the lymphatic veffels.

MANGROVE-TREE.

Of thefe there are feveral forts, and there is fcarce an inhabiter in Jamaica but knows them. The two moft noted are, the red and black mangroves. The roots of thefe mangroves are fo knit and entangled together in the water, and juft above the furface of the water, that they look like one continued tree for miles together; and it is to the roots of thefe trees, that are deep under water, that our oyfters ftick clofe, and grow together; from whence come the faying and notion, of oyfters growing on trees in Jamaica.

The bark of the red mangrove is made ufe of here for tanning, and does it to that perfection in fix weeks that oak-bark will not do in fix months time, and it is reckoned to give the moft lafting fole-leather in the world. It is a moft excellent reftringent: I have made a ftrong decoction of the mangrove-bark that would ftop bleeding, and dry up the great defluction of running ulcers. I had a fon that was extraordinarily full of the confluent fmall-pox, whofe foles of his feet feparated, and came off like the fole of a fhoe, and left his feet raw, and fo tender that he could not fet them upon the ground; upon which I fent for fome of the tanfat or liquor of this bark, fuch as they tan their leather with, and added a little alum, and boiled it up very

ftrong,

ſtrong, with which he bathed his feet every day; and in about a week's time, his feet were as hard and as firm as ever, and he was able to walk about without ſhoes on.

Another ſort hath a long black pod or fruit; and there is another, commonly called wild olive. Theſe are all of a binding and reſtringent quality, ſtopping all ſorts of fluxes.

MAPLE.

We have of the maple kind in America, as may be ſeen in Sir H. Sloane. The roots of maple, bruiſed and boiled with hogs fat, or *agnus caſtus* oil, applied as a poultice, take away the hardneſs of the ſpleen.

MARIGOLDS.

The garden marigold grows extraordinarily well with us; beſides which, we have many wild or field marigolds, ſome of which are exactly like the European wild marigolds, and are of the ſame nature. They are counted good againſt the yellow jaundice, and to diſcuſs impoſthumes; the roots bring away after-births; the flower made into a conſerve, with a little candied orange-peel, is a great cordial, comforting the heart and ſtrengthening the ſtomach; the diſtilled water, with ſugar of lead, cools inflamed and running ſore eyes, eaſing the hot pain of them.

MARSH-TREFOIL, or BUCKBANES.

We have many ſorts of trefoil, and alſo a marſh-trefoil, or a ſort of buckbane, growing in America: It hath a leaf like that of the water-lily, with a white flower. Marſh-trefoil, or buckbane, hath been of late much experienced in Pruſſia, to be a wonderful remedy in goutiſh diſtempers: They make a ſtrong decoction

coction of the leaves in ale, and drink a glass thereof every four hours during the paroxysm; from whence they find great relief. The decoction hath both a very disagreeable smell and taste, but the plant may be rendered more pleasant if prepared into a spirit, liquid extract, or syrup. Dr. Robinson recommends this plant as singularly useful in hydropic cases; and says, he observed scabby poor sheep, which have been put into marshes abounding with this herb, have soon recovered and been made fat by eating thereof; and that the Germans and other nations highly esteem it; that in all desperate diseases they have recourse to it, as a panacea, or universal remedy.

MASTICK.

There are in Jamaica three sorts of trees called mastick; *viz.* black, white, and yellow.
See Black, *&c.* Mastic.

MELONS.

Musk and water melons we have in great plenty. The seeds of melon and musk-melon are two of the four greater cold seeds, of which they make emulsions for the strangury occasioned by cantharides. They are great diuretics, and abate the heat of fevers and all inflammations.

MILK-WOOD

Is of the laurel-leaf kind. I have seen the boys in Jamaica get the milk of this tree, which immediately grows so tough and viscous, that they would put it upon twigs and branches of trees, by which they would catch parrots, parroquets, and several other birds, both great and small.

MILK-

MILK-WORT.

This is called blue Chili milk-wort. The natives make a cold infusion of this herb all night in water, and, drank in the morning, it proves a strong diuretic, and eafes pleuritic pains. It generally grows on mountainous land.

MINT.

Befides the common mint, which grows here in great plenty, we have alfo an herb which fmells like the Englifh cat-mint, and is of the fame nature, but ftronger. It is a fpecific to haften or facilitate labour, or the birth of the child; it expels wind, gives eafe in the cholic, and takes away cramps and convulfions occafioned by cold and moift humours falling upon the nerves; it alfo cures barrennefs in women.

MISLETOES.

The very fame fort that grows in England upon oaks, pear-trees, and fome others, grows in America upon dogwood, which is as hard as the Englifh oak, and of the fame virtue. It is good againft the falling ficknefs, is accounted a fpecific for moft difeafes of the head, and is one of the chief ingredients in the famous *pulvis epilepticus* of Riverius. The berries, bruifed and the juice expreffed from them, mixed with linfeed oil, and taken inwardly, cures pleurifies, ftitches and all pains of the fides, relieves palfies, convulfions, and cramps; made into a cataplafm, ripens fwellings and fchirrous tumours or impofthumes.

MOON-WORT

Is of the fern kind, and of the nature of *ofmundas,* which will be treated of hereafter.

MONEY-

Money-Wort,

Which fome call herb two-penny, grows in great plenty in America. They have the fame virtues with the Englifh money-wort, which is a fpecific for all forts of fluxes of the bowels, and is a good vulnerary. You may make an excellent balfam of the green herb, to heal wounds,

Mosses.

American moffes are much the fame with thofe of Europe, and of the fame virtues, differing in what they adhere to ; fome of which are under water ; fome fpreading upon the ground; fome fticking to trees, wood, and ftone ; and fome fticking to rocks that are conftantly wafhed with falt water. I had two forts brought me from a place called Wreck-Reef, of the fub-marine coralline kind.

The crufty mofs fpreads itfelf upon the rocks, after the manner of liver-wort, which the country people in Europe fcrape from the rocks, and, being finely pow- dered, they make a moift mafh, and put it into vef- fels fit for dying the cloth they intend, which it doth of a purple colour. John Francis Abela, in his de- fcription of Melita, mentions this mofs; which, he faith, is by the country people called *vercella*, which they dye wood with.

The other fort is *fucus marinus dictus roccella tincto-rum :* This makes a noble purple. Before the fucus is reduced into a tincture, the internal part is a whitifh red, and the external blackifh. The lively colour thereof is drawn out by maceration in urine, fo by lit- tle and little they gain the tincture ; and to fix it they ufe a little *fal alkali,* or *foda.* In making this colour, they put five times the quantity of the plant to one of

urine,

urine, or juſt as much as will macerate it, in-which it
lies a month; then they add a twelfth part of the *ſal
alkali*, or *ſoda*, to the macerated herb, which produ-
ceth a violet colour; this they heighten to purple, and
then to a ſcarlet or fine red, which ſome women uſe
as a waſh for the face, and is called *roccella*.

Mouse-Ear.

It is hot and dry, binding and conſolidating, and
therefore a good wound-herb.

Mug-Wort.

There is an herb in Jamaica called mug-wort, that
grows in all or moſt of the pooreſt grounds in America; nay, after a piece of ground is thrown up, being
worn out by planting, commonly the firſt weed that
appears is this. It is full of branches, which are covered with ſmall white flowers; its leaves are very much
jagged or ragged like rag-weed. In Jamaica, it is
called wild wormwood; the Spaniards call it *corbo
ſanta*. I ſaw, in the year 1723, a very great cure
performed upon a Jew, who, after a fever and ague,
had a violent inflammation and breaking out with ſores
on both his legs, which could not be cured by phyſic,
nor any ointment in the apothecaries ſhops; at laſt, he
was adviſed to *corbo ſanta*, to make a bath of it,
which he did, bathing twice a-day; and in three or
four days he was perfectly well, all his ſores healed
up, and the inflammation gone, with the great pain
that attended it. This I was an eye-witneſs to.

Mulliens

Are excellent wound-herbs, either inwardly or outwardly applied; they ſtop fluxes of the belly, help
ruptures, and are good in all coughs and ulcers in the
lungs, ſore mouths, and ulcerated piles.

Mushrooms.

MUSHROOMS.

There are three or four forts of mushrooms or fun-guffes in America, and but one that is fit to eat, which is the fame with thofe in Europe, and gathered and pickled in the fame manner.

Fungi albi venenati vifcidi. Thefe grow fo like the common inoffenfive fort, that feveral perfons have been deceived, and killed by eating of them. The fymptom is, that foon after they have eaten of them a hiccough feizes them, then a cold or chillinefs all over the body, attended with tremblings, and, at laft, con-vulfions, and death; for the circulation of the blood is ftopped. The antidotes againft it are, the *nahambu*, *jaborand, nhandiroba*, and fome other plants mentioned elfewhere.

The other moft venomous fort is one that rifes out of the ground about fix inches high, rounding, and hollow like a bladder, as red as fcarlet, full of holes like fine-wrought net-work.

MUSK-MALLOW.

Its ftalks are very hairy and rough; it hath a yellow flower, almoft as large and like the cotton-fhrub; its leaf is like the okra; its fruit is as big as the round okra, and hexangular. The feed of this plant fmells as fine as any mufk, and it is therefore called the mufk-mallow. The Egyptian women fet a great value upon it, for it helps barrennefs; it cures a ftinking breath, is a very great cordial, and expels wind. The feed is alfo called *abelmofch*, or *bamia mofchata*.

MUSK-WOOD.

This is vulgarly and commonly called alligator-wood. The bark of the tree is thin, of a whitifh-brown
without

without and reddish within, and of a most pleasant
scent, like musk. If you put a small piece of this
bark into a pipe of tobacco, and smoak it, it will perfume the room immediately. The wood also smells
like musk, as well as the bark; but as it grows old
and dry, its scent wears off.

MUSTARD.

Besides the common mustard plant, we have a wild
mustard, or a sort of Egyptian treacle-mustard. The
root of this plant is deep, large, white, and firmly
fixed in the ground by several smaller. The stalk is
very strong, round, hairy, and green, rising to about
four or five feet high, spreading branches on every side,
having fingered leaves standing on long foot-stalks. The
leaf is divided generally into seven parts or fingers;
they are viscid or clammy, will seem to stick to the
hand when you squeeze them, and have a rank disagreeable smell. The stalks and branches have short,
green, strong, straight prickles. The flowers come
out on every side of the tops of the branches: They
are each made up of four long petals of a white colour,
with some purple thrums or stamina. The pods are
small, round, and of a pale-green colour, inclosing a
great many very small brown seeds.

There is another sort, that hath a root four or five
inches long, small and white, with lateral fibres drawing its nourishment; the stalk is round, green, upright,
about two feet long, without any branches, having
leaves thinly placed thereon, without any order, standing three always together, on an inch foot-stalk, about
an inch and a half long and half an inch broad in the
middle; at the top of the stalk is a spike of tetrapetalous flowers mixed with purple, like the other sorts;
after which follows a three-inch long pod, small, round,

green,

green, like the other. The whole plant is balfamic and vulnerary: I have feen the very leaves applied to fores, and they would heal them; they give eafe in the gout; boiled in oil, remedy cutaneous difcafes, efpecially the leprofy. The leaves, boiled or decocted in water, expel poifon, provoke appetite, comfort the ftomach, caufe expectoration, and expel wind. The juice, with oil, helps deafnefs, dropped into the ear. The leaves, beaten and applied to the head, cure its aching from cold. Thefe grow in great plenty in all or moft parts of America, even in the worft and pooreft grounds, in yards, fides of the highways, and ftreets, without planting or cultivating.

MYRTLES.

Many kinds of myrtle grow in America; as the pie-mento, filver-wood, &c. All thefe are ever-greens; and one fort, viz. myrtus cotini folio, warmeth and ftrengtheneth the ftomach, expelling the wind, and eafing the cholic. A bath or fomentation of the leaves cleanfes and heals ulcers. All the myrtles are of a hot biting reftringent quality. There is a myrtle in North-America which affords a great quantity of green wax, of which they make candles in Carolina: I have feen great quantities brought to Jamaica, that burnt very well.

NAHAMBU, or NHAMBI.

It hath a fibrous root, from which arifes a pretty thick hard ftalk, knotty, rough, and hairy; fo are the branches. It hath a broad, juicy, green leaf, largely indented or divided, like the American celandine. From between the leaves come the flowers, on a long foot-ftalk, which are fingle and monopetalous; after which comes the fruit, which is round, and as big as

a little

a little cherry, covered over with a chefnut-like rough coat, in the fhape of a *ricinus;* in which are flat oval feeds, of a fhining yellowifh-brown colour. Every part of this plant hath a hot fpicy biting tafte upon the tongue, with an aromatic flavour. It is an excellent antidote againft all forts of cold poifons; for it is faid, that two or three drops of the juice of this plant, put upon a toad, immediately kills it. The powder of the bark, leaves, and fruit, expels the poifon of all other venomous creatures. In cholics and belly-aches, it eafes the pain and expels the wind.

Naseberry-Tree.

It is alfo called by fome *fappadillos;* but I take them to be of the mammee kind, having juft fuch an outfide as the *mammee fapota,* only they are much fmaller. The fruit is gathered when tree-ripe, but is not then fit to eat, being hard and milky, for a drop of milk comes out at the end where it joined to the foot-ftalk; but when they are laid up for two or three days, they grow foft and mellow, are of a very pleafant fweet tafte, and full of juice, like the Bergamot pear. In it there are two or three ftones or kernels, hard, black, and fhining as if polifhed, about the big-nefs and fhape of a prune-ftone.

Navel-Wort.

We have in America a water navel-wort, that grows in ditches and moft ftanding waters in great plenty. It hath a fmall round root, under the furface of the earth; at the joints are a great many fmall hairy blackifh fibres, by which the plant is nourifhed; and from the fame places are fent up the leaves and flowers, upon pretty long foot-ftalks. The leaves are round, thick, fnuated on the edges, fmooth, above an inch diame-

ter,

ter, and very green, the foot-ftalk entering in their very centre. The flowers ftand clofe togeter round their foot-ftalk's end; they are many, joined together, and of a greenifh colour. The feeds are broad like parfnip-feed. The plant is fharp to the tafte, and has been taken by fome planters for fcurvy-grafs; the whole plant is of hot and fubtle parts, pleafant and aromatic to the tafte: They open obftructions of the liver and reins, for which no remedy is more proper; the juice of the green leaves is a famous antidote againft poifon; and the native Brafilians procure vomiting with it. It is ufed to take away the fpots which the Portuguefe call *os figados*, which are liver-fpots; and it is faid to kill fheep, if they feed upon it.

Nephritic-Tree.

This tree is fo called in Jamaica for its being a fovereign remedy for the ftone, gravel, and difficulty of making urine; it is alfo good in obftructions of the liver and fpleen. The ufe of it was difcovered to our traders to the main continent of America, where a Spanifh bifhop did fuch wonders with it for the gravel and ftone, that, being willing it fhould be known for a public benefit of mankind, he fhewed the fhrub or tree to fome of our merchants, who foon found the fame tree in Jamaica, but chiefly about St. Jago de la Vega, for which reafon it is believed the Spaniards planted them; for if you go above four or five miles from that town, you will hardly meet with one of thefe trees throughout the ifland. It has a moffy flower, that fmells as fweet as the Englifh May or hawthorn; is a large fhrub, with little roundifh leaves; the whole plant grows almoft like an Englifh maple, but is full of fmall prickles; its leaves glaffy, fmall, and round; its flowers are like the fingrigo; its fruit is a fmall long

red

red pod, which when ripe opens of itfelf, turning infide out, curling, and twifting, fhewing a black bean,
with a white poppy down fubftance at one end, in the
fhape of a kidney. Upon this account, faid the Spanifh bifhop, nature points out the ufe of this plant;
the bean itfelf is in fhape of the kidney, and that white
poppy fubftance about it fignifies the fat of the kidney.
It is the bark which is chiefly ufed: When decocted,
it fmells like new wort, but a little bitterifh; of which
they muft drink plentifully; it worketh by urine. I
have often given it with good fuccefs; but I am of
opinion the fruit would be found to be prevalent if
experienced; for the bark is fo ufed, that it is now
rare to meet with a tree that hath not been barked.

NETTLES.

There are many forts of nettles growing in America,
and fome of them more ftinging than any in England.
I take the American nettles to have the fame virtues
as thofe of England. The ftinging fort is good againft
tympanies or dropfies, occafioned by a ftoppage of
urine: The juice of the leaves is good for thofe that
evacuate a vifcid or purulent urine, which negroes are
very fubject to; and, mixed with fugar, milk, and a
little flour of brimftone, drives out and cures the itch.
Thofe that do not fting are much of the fame nature
of thofe that do; for thofe that fting, do it not by any
different heat of the plant, but by their downy or hairy
prickles being harder and ftiffer, piercing into the
fkin like points of needles; and when that fharpnefs is
taken away, either by the fire, or the heat of the fun,
thofe nettles fting no more than dead nettles, which
are good pectoral herbs, &c.

NHAN-

NHANDIROBA, *or* GHANDIROBA.

The firſt time I met with this plant was in St. Thomas in the Vale, in that part called Sixteen-Mile Walk, in Jamaica; where I ſaw it climbing and running up to the tops of very high trees. It happened to have its fruit upon it: Its leaf very much reſembles the Engliſh ivy-leaf; but its fruit is like a green calabaſh, only it has a circular black line round it, and two or three warts, or little knobs; the inſide of the ſhell is full of white flattiſh beans, incloſed in a white membranous ſubſtance; and, when thorough ripe, the fruit turns browniſh as a ripe calabaſh, and the beans or nuts are then of a lightiſh-brown colour, and have a thin hard cruſt, in which is a whitiſh kernel, full of oil, and exceſſive bitter. The nuts or beans, which are generally ten or twelve in a ſhell, are ſo cloſe and compreſſed, that when I have taken them out, I never could place them ſo again as to make the ſhell contain them.

Piſo ſaith, that he has ſeen whole families in Brazil, that have had violent aches and pains, got by the night-air, who have been cured with the oil of theſe nuts, which they may eaſily have growing in great plenty in moſt parts of America. It cannot be uſed in victuals, being ſo exceſſive bitter. A French gentleman, ſome years paſt, brought me from Peru ſome of theſe nuts, and aſked me, if I knew what they were? I did not ſatisfy him whether I knew them, but aſked him what the Spaniards called them, and what uſe they put them to? He told me, that the Spaniards called them *avilla*; and that they were worth their weight in gold to expel poiſon, and wiſhed I could find them growing in Jamaica; which they do in great plenty, and the negroes that I employed to get them for me called them *ſabo*.

NICKERS.

NICKERS.

There are two forts of thefe trees which are called nickers, the boys playing with the cone or fruit as they do with marbles: The one hath a yellow cone, the other an afh-coloured one. Its prickies are fhort and crooked, as the cockfpur-tree is; it hath a long fpike, full of yellow flowers; the pods or hufks are full of rough prickles, like the chefnut, but fharper, and fo ftiff as to prick the finger if you touch them; within this rough pod or cafe are four or five hard cones, which are called nickers, fo hard that the teeth cannot crack them. The Indians and negroes make ufe of them in venereal cafes, and fay they purge and carry off the caufe, and afterwards bind and ftrengthen the part. They grow alfo in the Eaftern parts of the world; for the Egyptians, in Alexandria, account them a fort of guard for their children againft witchcraft and forcery, hanging them about their necks as amulets. The fruit, finely pulverized, and given half a drachm, helpeth the meagrim, the torture of drawing the mouth of one fide, as alfo convulfions, and falling ficknefs.

NIGHTSHADES.

There is great variety of nightfhades in America, exceeding in number thofe in Europe.

1. *Solanum bacciferum Americanum flore corymbofo.* Sir H. Sloane makes it a valerian with a chickweed leaf. It grows very common in moft parts of America, and feems to be a cold and moift herb; but I do not remember I ever faw any thing eat of it, or that it is of any great ufe in phyfic.

2. *Solanum racemofum Americanum.* It hath a large round reddifh ftalk of the thicknefs of one's thumb, rifing four or five feet high, fet without order, with

many

many very large leaves full of veins, fome greater and fome fmaller. From the joints where the leaves ftand come forth feveral fmall ftalks, with flowers of a pale red, confifting of four leaves, ftanding in clufters, which bring forth fmall blackifh round feeds, four in a hufk or capfula. The root of this plant is very white and large, like a briony, and above a foot long; generally the ftalks of thefe are as red as an amaranthus, which makes the Englifh in Virginia call it red-weed; and the Indians in New-England dye their fkins with it, and the barks wherewith they make their bafkets. This night-fhade is a familiar purge in Virginia and New-England; a fpoonful or two of the juice of the root worketh ftrongly, and fo doth the extract; but when the root is dry, it lofeth its purging quality. I have known negroes in Jamaica who have taken them for a wild yam, and have eat them as fuch, which made them very fick, and purged them ftrongly.

3. This is the *folanum tuberofum efculentum*, or Virginia potatoe, whofe ftalk is two or three cubits high, fometimes five or fix, and is an inch thick, round, juicy, and channelled, fomewhat hairy, of a green colour, marked with many reddifh fpots, hollow and branched: The branches are weak, and, if not propped, lie flat upon the ground. It has feveral leaves fet by pairs upon the fame rib; three, fometimes four or more pairs, join in the compofition of one; but one leaf unpaired is greater than the reft. The leaves are fomewhat hairy, of a dark-green colour, and fhining upon the upper fide, but underneath neither of fo deep a green nor fhining: Between each pair grow other intermediate leaves, little and round, which make up the compofition above mentioned: The flowers are equal in bignefs to thofe of the wild mallow; they fmell like the lime-tree flowers: Thefe flowers are fucceeded by

an

an equal number of little apples, about the bignefs of
a chefnut, but of an orbicular form (like thofe of the
feed-yam in Jamaica), at firft of a dark-green colour,
but when ripe of a dark-red: They are full of a moift
whitifh pulp, in which lie many fmall roundifh feeds,
like thofe of the nightfhades, or what we call the hog
or red-pop in Jamaica. The root is tuberous, about
the bignefs of a man's fift, and from five to eight or
nine inches long. At the origin of the ftalk are many
fibrous roots, to which adhere other little fmall tuberous
roots; fo that the plant, when digged out of the earth,
will have fometimes fifty knobs of different magnitude,
juft like the white feed-yam in Jamaica; thefe knobs
we plant again for increafe. This plant was firft
brought from Virginia to England, and from thence
carried into France and other countries. In Virginia
the roots are called *openanck;* they eat it boiled and
roafted, as we do yams or potatoes: The Indians
make a fort of bread of them they call *chunno;* they
alfo flice the roots and dry them in the fun, and then
beat and fift them into flour: It is reckoned good and
wholefome nourifhment. From the frefh roots of this
plant the natives make a drink which ferments, and is
called *mobby,* or *jetici,* which they fuddle and get drunk
with, as they do with potatoe mobby in Jamaica.

4. *Solanum racemofum Americanum minus.* This
has a fmall oblong fibrous root, which fends up one
green round ftalk, two feet high, having many branches.
The leaves ftand on the ftalks without any order. The
flowers come to a fpike on the tops of the branches,
which have fome large hairs, or foft prickles; they
have a very fhort foot-ftalk. The flowers are white and
tetrapetalous, or four-leaved; after which follow fome
fmall berries, at firft green, and then red. Thefe grow
in all or moft iflands, and upon the main continent of
America.

5. *Solanum*

5. *Solanum bacciferum*, feu *officinarum*. This has a green ſtem, as big as one's little finger, riſing two or three feet high, the branches ſpreading themſelves on every ſide; the leaves are about an inch and a half long, and half as broad in the middle, where it is broadeſt, ſtanding upon a very ſhort foot-ſtalk; they are ſoft, of a dark-green colour, and jagged on the edges. To-wards the tops of the bunches come the flowers, ſeveral together, upon a ſhort foot-ſtalk; each flower is made up with five white or pale-yellow leaves, with orange-colour apices, ſtanding up in the middle of the flower, making an *umbo*. After theſe follow round berries as big as Engliſh peaſe, ſmooth, and black when ripe, con-taining a thin greeniſh pulp, with a great many round flat white ſeeds. I was ſurpriſed to ſee the Angola negroes eat it as coilu, or as we do ſpinage, without any prejudice, being ſo like the deadly night-ſhade in Europe The bark of this plant, bruiſed and put into water, intoxicates fiſh, ſo that they may be eaſily taken, but doth not kill them. The leaves are reckoned cooling, reſtringent, and anodyne; the juice, being put up the *anus*, eaſes pain and abates inflammation, and it doth ſo in *eryſipelas*, or St. Anthony's fire; but it ought to be cautiouſly uſed, being very cooling and reſtringent, and therefore too repercuſſive or repelling. The juice I know to be good in cancerous tumours and inflammations, and the diſtilled water is good in fevers. The leaves, juice, or oil, applied to the head, is good in frenzies from heat, and for inflammations, and fiſſures or cracks of the nipples of the breaſt.

6. *Solanum bacciferum, caule et foliis tomento-inca-nis ſpinoſis flore luteo fructu croceo minore.* This grows very common every where, even about the ſtreets of towns and villages. The ſtalks are very thick ſet with ſhort crooked prickles, the points downwards, woolly,

round, and about three or four feet high; the leaves are pretty large, and deeply finuated on the edges, and its big rib is fet underneath with finall prickles, fo that they make a good fence; the flowers are monopetalous, though the *ora* be divided into five petala, reflected back, of a yellowifh colour, with apices like the reft of the folanums; then come round orange-coloured berries, as big as Englifh peafe, having five green capfula under them; the berries are full of an orange-coloured pulp, containing finall white feeds. Their roots are very bitter, and of thin parts, and excellent virtue, efpecially the male: Half an ounce, in powder, purges all humours downwards, opens obftructions of the liver and proftrates, provoking urine, being ufed inftead of the opening roots, which are fo much efteemed. The decoction of the roots is diuretic, and good in burning fevers, and with honey in catarrhs, and in the ftrangury, with fome cardamoms it expels wind. The decoction of the leaves, with fugar and limes, is good for the itch. The juice of the roots and leaves is good for confumptions, and with fugar for the forenefs of the breaft.

7. *Solanum fruticofum bacciferum fpinofum flore cœruleo.* This grows like the former, but its leaves and fruit are like thofe of *amomum Plinii.* The flowers of this are blue, and the berries red.

There are alfo,

1. The fhrubby nightfhade, with a branching leaf.

2. The fhrubby prickly nightfhade, with peach-tree leaves.

3. The fhrubby and prickly nightfhade, with laurel leaves.

4. The climbing nightfhade, with woolly leaves.

5. The woolly nightfhade, with a mullien leaf, and finall yellow berries.

6. Tree

6. Tree nightfhade, with a leaf like the common garden nightfhade, with a fmall fcarlet-coloured berry.

7. Tree nightfhade, with an undulated or waved almond-tree leaf, with a large white flower and red fruit.

8. The climbing nightfhade, with a henbane leaf, and a flower with a purple colour without, and white within.

They have all much the fame nature and quality with the feven forts mentioned above.

Befides thefe nightfhades already mentioned, Father Feuillee takes notice of two forts growing in Chili; the one oak-leaved, bitter fweet, with purple flowers; the other the *folanum chenopodioides acinis albefcentibus.* The natives were ignorant of the virtue of this plant until the negroes came amongft them, who were fubject to a certain difeafe which killed them in their prime: It was an extraordinary extenfion of the *anus,* attended with a fever, which was fo mortal that many of them died before they difcovered the remedy. They take the juice of the tops of this plant, mixing it with rofe-water and a little alum, which they apply to the part, and a little taken inwardly cures the diftemper. The fame, being applied to the eyes, takes away inflammation, pain, and dimnefs. This fovereign plant grows about a yard high, about the mountains of Valparaifo, and many other parts of South-America.

Oak of Cappadocia.

It hath a ftrong, ftriated, woody, folid ftem, as big as one's little finger, growing about three or four feet high. Its leaves are cut and divided juft as mugwort leaves, but are a little larger, of a very dark-green colour above, but underneath more pale; and upon the top twig come out a great many fmall mufcous flowers, of a yellow colour, fet clofe together as in

H 4   others

others of this kind. The fruit is an echinated or rough husk, just like the fruit of *tribulus ;* and the seed is like grape-seed. The whole plant has a very strong smell, like the others of this kind. There is a notion of this herb, that if it be put under the sick's pillow, it foretels death if he sleep not. Boiled in *cergilim,* that is, sesamum and burnt wine, and applied to the part affected, it cures empyemas and abscesses of the stomach, before they ripen, especially if the juice be drank with honey; made into a plaister with horehound, it cures the cramp or spasm; with honey, eaten fasting, it cures the dropsy. The root, boiled in the above-said oil, takes out freckles or spots; boiled with cocoa-nut milk, it cures ulcers, and so doth the bark, powdered and sprinkled upon them; it eases after-pains.

OIL-NUTS.

These are so called from the great quantity of oil got out of them; and also vulgarly, but very erroneously, called *agnus castus,* they having no relation to that species; but every body in Jamaica calls it *agnus castus,* or oil-leaves, which they put to their blisters instead of melilot, and use no other. The root, decocted and drank, cures the cholic and swelling of the belly and legs; and so doth the leaves, boiled with wild ginger and ground-ivy, and then fermented with a little sugar or melasses, which will purge very strongly. Planters have not only cured dropsies in negroes with this drink, but also the yaws and venereal complaints, taking away the gummous nodes, and pains in the joints. The leaves, applied to the head in fevers, remove pain; a cataplasm made of the green leaves, cassada flour, and a little oil of the nuts, applied to womens breasts, softens and discusses the coagulated milk and hardness; and, if not to be discussed, it will ripen it, bring it to digestion, and break it.

Negroes

Negroes are troubled with a diftemper in their legs, which they call a guinea-worm : The firft appearance is a hard fwelling, with much pain and inflammation; and fome time after will appear, through the flefh and fkin, the head of the worm, as fmall as a knitting-needle, which they take hold of, and draw it a little, and get it round the quilly part of a fmall feather; but if they draw it fo hard as to break it, many ill accidents will attend the part, and fometimes gangrenes enfue. Now, to ripen and forward the work, make a poultice as before directed, and lay over it one of the leaves, which will foften and bring the worm out, by turning the feather every day, drawing a little at a time, and by degrees the worm will come entirely out, which fometimes will be feveral yards long, and not bigger than a thread; fometimes, barely anointing the part with the oil, and laying a leaf upon it, will do. The oil of this nut purges ftrongly; and I knew one that would boldly give an ounce or an ounce and an half, in what they call the dry belly-ache, which would go through the patient when nothing elfe would; outwardly, it is good for cold aches and pains, or cramps and contractions. Its oil will keep without being fetid or ftinking, and therefore may be converted to feveral ufes.

OILY PULSE,

Which is called *zefamum*, or *fefamum Africanum.* The firft time I faw this plant, it was growing in a negro's plantation, who told me, they ground the feed between two ftones, and eat it as they do corn. I obferved it hath a fmall long fibrous root, from whence fprings up a ftraight fquare ftalk, like a nettle, two or three feet high, fet about with long leaves oppofite to one another, and jagged, much refembling the *lamium,* or archangel; and at the tops of the ftalks come forth divers white flowers, like *digitalis*; after which come their feed-veffels, full of fmall white feeds, which the negroes call *foonga*, or *wolongo*, which is much like the fago fold in fhops, but very oily. The oil that is drawn from it is called *crrg lim* oil. The feed is often mixed and ground with coco, to make chocolate. In Ethiopia and Egypt, they ufe the oil as we do oil-olive: It is made by grinding the feed, and expreffing the oil, as they do by other feeds. The feed and oil are hot, moift, emollient, and refolving; breed grofs nourifhment, and therefore hurtful to weak ftomachs. Dropped into the ear, it is good to foften the hard wax, and help deafnefs. A decoction of the plant is good for coughs, pleurifies, inflammations of the lungs, hard fchirrous tumours, and women ufe it for hardnefs of the womb. The herb and feed, boiled in honey, make a good cataplafm or poultice for hard tumours, and dried nerves or fhrunk finews; fo doth the oil. A decoction of the whole herb, flowers, and feeds, is good in clyfters, to foften the belly, and give a ftool or two. The juice of the herb or diftilled water is good for fore eyes. The decocted feed fattens, the oil more, and the dregs (which are eaten for food in Ethiopia) more than the oil; women often drink the oil, to be fat. The dregs (when they make the oil by boiling) is given to four ounces in pleurifies and pains, and in all difeafes of the fkin, outwardly as well as inwardly. In Greece, they ufe it for cakes, inixing it in making their bread. In Bengal it is planted to make oil; but it makes ground poor. The oil takes off the roughnefs of the throat, clears the voice, and mollifies hard impofthumes. This oil is better for making odoriferous oils than others, becaufe of its durability. The oil, if taken to four ounces for many

days,

days, is good againſt the itch, hard breathing, pleu‑
riſies, pains in the ſtomach, womb, and guts, and is
every way as effectual as linſeed oil. Sir H. Sloane
ſaith, that Mr. James Cunningham, F. R. S. and his
very good friend, wrote to him from China, where he
was phyſician to the Engliſh factory, informing him,
that the bean, or mandarin broth, ſo frequently men‑
tioned in the Dutch Embaſſy, and other authors, is
only an emulſion made of the ſeeds of ſeſamum and
hot water.

OKRA

Is of the mallow kind. The fruit, when green, is
cut croſs-ways with its ſeeds, dried, and ſent to Eng‑
land and other parts of Europe, to make their rich
ſoups. Thoſe that frequent Pontack's have often eat
of it, paid well for it, and knew not what they were
eating at the ſame time.

They are very cooling, emollient, and of great nou‑
riſhment; very proper for diſeaſes of the breaſt, and
provoke urine, ſtone, and gravel, having all the virtue
of the marſh-mallows. I adviſed a perſon that was in
a deep conſumption, and of a depraved appetite, of a
cadaverous countenance, and a mere ſkeleton, to have
always the dried ſeed of the okras by him, that he might
not be without them all the year round; the which I
ordered him to have beat into a fine flour, ſeparating
the huſks from it, and ſo to thicken all his broths or
ſoups with this flour; which afforded him ſo much
nouriſhment, taking away his hectic fever, that, in leſs
than twelve months, he was as ſtrong and luſty as
ever he was all his life-time, and gave me many thanks
for my advice.

OLD.

OLD MENS BEARD.

It is a ftrange and uncommon name to give to any thing of the vegetable kind; but their great refemblance to a whitifh hoar, hanging down, makes it fo called. It is no more than the vifcus of a tree; it hangs down upon branches of trees like hair, but chiefly upon ebonies and manchioneel trees, of whitifh colour. Dried and beaten very well, it makes good ftuffing for faddles, or to pack up any thing, as well as tow or flax. It is of a drying, binding quality.

OLEANDER, *or* ROSE-BAY.

I met with a very fine beautiful oleander, with double carnation flowers, like a rofe, but not fo fpreading. They that had of it told me, they had the feed from Peru, and called it rofe of Jericho; but I told them that was a great miftake, for it was an oleander with double flowers, having the fame fort of leaf. It is of little or no ufe in phyfic. Some affirm it is venomous, taken inwardly.

OLIVES.

In fome parts of the main continent of America, they have of the tree-olive, as good and as large as in any part of the world; in Jamaica, they have not got them; not but that they would grow admirably well there, as we fee by the wild olives, which grow in great plenty. Of thefe there are two or three forts: One is made ufe of for green walks, and hath a fruit like the luke or Lucca olive. Another fort, they call olive-trees, are very large, tall, fpreading trees, whofe wood is excellent timber; and its bark is made ufe of to tan leather, mixing it with mangrove bark. I have made an excellent reftringent ftyptic water of the
bark.

bark. The bully-tree, aforementioned, bears a very exact olive, which might be improved.

ONAGRA

Is a fort of yellow-flowered loofe-ftrife, or rofe-bay willow herb.

ONOBRYCHIS, *or* COCK's HEAD.

The *onobrychis,* in America, feems to be more of the *hedyfarum* kind, or hatchet vetch; of which there are feveral forts:

1. *Hedyfarum triphyllum fruticofum flore purpureo filiqua varie diftorta:* This hath a woody brown-coloured ftem, having feveral green rough branches, four feet high. The leaves come out on every fide, without any order, three always together upon a ftalk, fmooth above, of a dark-green colour, and rough underneath; the tops are long fpikes of flowers, papilionaceous, of a pale purple colour; after thefe follow feveral pods, flender, rough, jointed, and varioufly turned and diftorted. The plant purgeth a little; for if an ounce of the dried leaves be put in a purging decoction, it furthereth the purging property, caufing not only watery humours to be voided, but thofe that are tough and clammy; alfo, it helps to digeft cold humours.

2. *Hedyfarum triphyllum fruticofum minus.* This grows much like the other.

3. *Hedyfarum triphyllum fruticofum flore purpureo.* This grows like the former. The root of this is hot, and a decoction of it, in water or other vehicle, is one of the beft remedies againft cold fluxes of the belly. The fume or fmoak of the leaves, received with the head covered, cures the head-ache which comes from cold. All the forts of *hedyfara,* efpecially the feeds, are bitter, and therefore good ftomachics and expellers

lers of poifons.   They open obftructions, and kill worms.

Opuntia.

This is an American name for what fome call prickly pears, of which there are feveral forts.

1. The common prickly pear.

2. Another kind, whofe flowers are of a beautiful red.   It has a fucculent juicy leaf, but no prickle, nor has its fruit.   Some call it the true cochineal, as if its feed or flower was the cochineal ; but that is a vulgar error : This fhrub is only the food for cochineal, which is an infect or reptile.   Many or moft that have touched or difcourfed upon cochineal, have fallen into miftake, taking the plant they feed upon for the cochineal, when the cochineal is an infect that feeds upon this plant, and the goodnefs of it is owing to their feeding ; for we have enough of the fpecies fticking to feveral plants in Jamaica, but thofe plants not being their proper food, they have little or no red tincture in them.   From *Tlaxcala*, a city in Mexico, they deal for 2,0,000 crowns worth a-year.   There are four forts of cochineal : 1. Is called *tufkaliobe*, which is of a black dull colour, but the longeft grain ; 2. Is *mifteka*, which is a grey fort, and worft of all ; 3. Is *guaxaca*, in colour between both, and of the fame fize, but much excels the others in goodnefs ; the 4th fort, which is the *tlaxcalla*, or *rofella*, which is the reddeft and richeft of all : But the merchants, for covetoufnefs, generally mix all together.   Choofe that which is plump, large, well fed, clean, dry, of a filver colour on the outfide, and, when chewed, tinges the fpittle of a bright-red colour ; reject that which is meagre, falt, and light, and take care there be not fmall fand, grit, or ftones, in the infide, which will make it weigh, and enhance the price.

Oranges.

ORANGES.

In America there grow oranges of all forts in great plenty, and as good as in any part of the world, and fome as bad, for there are both fweet and four, bitter and infipid. They are fo well known that they need no particular defcription, and therefore we fhall treat more largely of their virtues.

Orange-peels are oily, bitter, and hot, and therefore warm and comfort a cold ftomach, expel wind, and help digeftion; chewed and fwallowed upon an empty ftomach, they prevent the cholic. My father, who was an experienced phyfician, made a conferve of the peels of fweet or china oranges, which he adminiftered in cold vifcous humours of the lungs, and in that which fome call rifing of the lights, great fpitting, and flimy matter in the glands, with good fuccefs. If the flowers were added to it, it would make it more prevalent. The famous Boyle faith, orange-peels cure the cholic; and Etmuller faith, they provoke urine.

The effence is a fpecific in the cholic; the preferved peel is a good ftomachic. Five ounces of the juice taken at a time, drive forth putrid humours by fweat, and fortify the heart. The diftilled water of the flowers is very odoriferous, and is good againft contagious and peftilential fevers; it alfo helps cold and moift infirmities of the womb. The butter or ointment made of the flowers, and mixed with a little of its effence, is excellent to anoint childrens ftomachs and bellies, comforts and warms the ftomach and bowels, eafes the gripes, and kills worms.

ORTICIA

Is a plant that grows in Chili, and is a fort of ftinging *palma Chrifti*. It is a violent emetic and cathartic.

OSMUNDAS.

OSMUNDAS

Are of the fern kind. The only difference that I fee between ofmundas and common ferns, is in their growing upright without branching, and both ftalks, and under the pennas, are full of ferruginous duft, &c. They have the fame virtues as common ferns; befides which, they are accounted fpecifics for rickets in children. A decoction of them, drank plentifully, forwards the healing of wounds, ulcers, &c.

OYSTER-GREEN

Is a fub-marine plant; fome call it *flanke*. It is of the nature of other fea-weeds, which is cooling, drying, and binding; is good againft inflammations and the hot gout, and is faid to kill worms.

PAICA JULLA.

This grows about Lima and Callao. Its flower-rim is white, and is compofed of fix yellow petals. It is a purging plant, but rarely ufed, by reafon of its violence. They alfo think it a poifon, becaufe it kills a houfe-animal, called *cueiz* in Peru and Chili, in Jamaica called wood-flave; and therefore it is called *cueiz*-bane.

PAJOMIRIOBA.

There are two forts of this plant. The firft fort hath a dark-greenifh woody ftalk, rifing from fibrous roots about three feet high, having many fmall ftalks coming out on each fide; and upon each ftalk come out eight or nine leaves, without any manner of foot-ftalk, oppofite to one another, about two inches long, and half an inch broad where broadeft, which is towards the ftalk, and then goes off tapering with a fharp point;

at

at the end of the branches come out its flowers, which are pentapetalous, and of a yellow colour; after the flowers come fmall flat flender pods, from four to fix inches long, which, when ripe, grow brown, and open; their feeds are a little bigger than lentils. It flowers and bears feed all the year.

The fecond fort grows much like the former in moft refpects, only is a little fmaller, and the leaves round inflead of being pointed at the ends. The root is powerful againft poifon; the feed, bruifed and mixed with vinegar, prevails againft ring-worms. The whole plant is cooling and cleanfing, and therefore good in ulcers; fteeped as you do indigo, it will afford a black-ifh-blue muddy fubftance, which is excellent for the galled back of a horfe, and other fores. It is called by fome, wild indigo.

Palghi

Is the name that the South-American Indians give to a fort of fmall fage, which grows up to a bufh. The leaf fomewhat refembles rofemary, or what they call wild rofemary in Jamaica. It fmells like Hungary water, and muft contain much volatility, if we may judge by the fcent and tafte.

Palqui

Is the name the Indians give to a fort of very ftink-ing wall-wort, having a yellow flower like it, which cures fcald-heads and fcurf.

Palms.

1. The date-tree. The unripe dates are very harfh and binding, and the ripe alfo while they are frefh, but not fo when they are dry. They ftop vomiting and fluxes, and check the menftrual difcharge; they are

I   alfo

alfo proper for relaxation of the fundament and piles, being taken in red wine.

2. The palm-tree, from which the oil and wine are got. It is from the fruit that they get oil; when they are thorough ripe, there is, between the outward fkin and the ftone, a yellow pulpy fweet fubftance; this pulp turns to a thick oil, like butter, as it grows old, and of a reddifh-yellow colour; alfo, the inward kernel turns to oil in the fame manner. It is an excellent fuppling oil; the traders for flaves, when they expofe them for fale, fhave them very clofe, and then anoint their bodies, limbs, and joints with it, which makes them look fmooth, fleek, and young. From the body of the tree by tapping, and the branches before they have fruit, they get a liquor which is called palm-wine, and fo ftrong as will inebriate or caufe drunkennefs.

3. The palm from which they get the cabbage, which is only the green top, that is about a yard long, the outward parts being taken off, which are thick foldings or coats, one over the other, until you come near the centre or moft inward parts, which is as white as fnow, and that which breaks or fnaps fhort without ftrings is good cabbage. I obferved, that after the firft coat is pulled off, which is a very green colour without-fide, the infide is very white, and fo are all the reft until you come to the cabbage, and the nearer you approach to it, every tunicle or coat grows thinner; and perhaps there are five or fix of thefe coats or fkins before you come at the good cabbage. I alfo obferved, that thefe fkins are finer and whiter than paper, and with a ftylus or fteel pencil you may write any thing you have a mind, which is not to be rubbed out, but as lafting as the leaf itfelf, which may be dried and kept for ever in what fhape you pleafe. The trunk of
this

this tree is very fmooth and ftraight all the way to the top, which is fometimes fifty or fixty feet; but when they are fo tall and old, their cabbage is not good; one of about fifteen or fixteen feet high, and which looks very green at top, produces good cabbage, and in great quantity. From the top fpring twigs or fmall branches, full of fmall flowers; and then follow fmall round berries, of the bignefs of an hazel-nut, which the birds eat and mute the ftone, by which there is a continual fupply of them, otherwife they would foon be deftroyed; for when one is cut down, there is never any fpring from the root again; or if the top be broke off this, or any of the palm kind, they never grow again. The outfide of this tree is fo hard that a bullet will hardly enter into it, but it is not an inch thick; the reft, within-fide, is nothing but a foft pappy fubftance. The Spaniards cafed their houfes with boards of thefe, which were found to ftand firmer than any other houfe againft earthquakes and hurricanes.

4. The coco, or coker, or coco-nut tree. This is the largeft, in general, of all the palms; for although the cabbage-tree fometimes, in open ground, and thofe very old, grow to be forty or fifty feet, yet in general they are feldom above twenty feet high; whereas the coco-tree generally grows to forty, fifty, or fixty feet high, and, if no accident happens to break its top, will ftand fixty, feventy, or an hundred years. They are fmooth and without any prickles, having no branches but towards the top. Their ftalks, with its leaves, are like large limbs of trees, one ftalk being as big as a man's arm, and ten or twelve feet long, befet with leaves on each fide, long and narrow, and not above an inch broad. Near the top come out many branches or twigs, upon which the fruit grows, which is very

I 2   large

large and green, about a foot long, weighing five or six pounds weight. All the fubftance of this fruit, from its outer part to the fhell, is made up of a tough thready fubftance, of which is not only made cordage and tackle for fhips, but caulking ftuff, which is better to caulk with than oakum; and being fteeped in water, and beaten as flax-weed, makes excellent cloth for feveral ufes. After this thready fubftance is taken off, there appears a large hard fhell, having at the head or top three holes, and a little protuberance between, which fomewhat refembles the nofe and eyes of a monkey: Thefe fhells, being polifhed, not only make cups to drink in, but alfo are fet in filver for ornament, and feveral other ufes. Within the fhell is a very white fubftance, about half an inch thick, adhering clofe to the fhell, which is firm and hard, tafting like an almond while it is frefh gathered; but, feraped out and put in the fun, it turns yellow and oily, or fat like butter or like palm-oil, and of the fame ufe; but it will not keep long. The reft of the cavity of this fhell is filled up with a fine, clear, fweet, cooling liquor, as pleafant as milk; which will not keep long out of the fhell, foon turning four like vinegar; but, in the fhell, the liquor will all become a perfect kernel in about twelve months time, if you keep the fruit with its outward bark upon it (otherwife it will not do fo). Of this kernel are made fine fweetmeats. They alfo draw a liquor from this tree, either by cutting the branches that bear the fruit (to which they faften veffels to receive the liquor), or by boring the body and plugging it, after which they let out the liquor when and how they think fit; this liquor they call *fum*. It taftes like new fweet wine; this they fometimes boil up into a grain like fugar, which they call *jagra*. If you expofe the liquor in the fun, it will foon

turn

turn to vinegar; but, diftilled in its ferment, it makes a fpirit called *orraqua*, or *rack*, which far exceeds that made from rice; and thefe trees being called in fome places *toddie*, it is therefore called toddie-rack. The frefh meat or kernel of this fruit is of great nourifh-ment, therefore good in emaciated bodies; it is faid to be a great provocative, and is good to take away the roughnefs and hoarfenefs of the voice. But the Americans, not knowing the great ufes of this tree, do not fet fuch a value upon it as thofe in the Eaft In-dies; for there cannot be found in the whole world a tree that hath fo many neceffaries for the ufe of man-kind; and it may properly be faid of it, that it affurd-eth meat, drink, and cloathing.

5. The palmeto-royal, which makes the beft cover-ing for houfes.

6. The little round thatch, which grows more com-mon than palmeto-royal, and more made ufe of for covering houfes.

7. The great macaw-tree, already defcribed.

8. The fmall macaw-tree.

9. The prickly-pole. It beareth a fmall round red berry, which pigeons feed on; it hath a fweet yel-low pulp, between the outward red fkin and the ftone. It is with this prickly palm that the Indians arm their arrows, being as hard as iron: The arrow itfelf is the flag of a fugar or wild cane, that grows out of the middle and top of the cane, being light, ftraight, and fmooth as a dragon-blood cane. Of this they take about four or five feet, and, at the end, they put a fmall fharp fpike, of about a foot long, of this prickly palm, in which they make nicks to lay their poifon in, and beard it to hinder its being drawn out from the wounded part.

10. The large broad round thatch. It is fuppofed

I 3

the

the Spaniards in America get from this the gum called *caranna*, which being of value, they endeavour to conceal it. It is a very large-bodied tree, rather bigger than any other palms. I have seen several hundred of them growing in one small savanna. They are about thirty or forty feet high, and have a large branching top, with very thick stalks, as thick as a man's wrist; at the end of which is a broad spreading palm, which when cut into a fan towards the stalk, will be above a foot over, and make a semicircle of above two feet; this they stain or dye of several colours, making commodious fans to fan people, and keep off the flies while they sleep. The leaves they blanch, and make fine *bongraces* and hats of, &c.

PANKE

Grows chiefly in the kingdom of Chili, although it is to be found in most parts of South America. There are two sorts; the one, they eat the raw stalks of peeled, which are of a sweetish agreeable taste; they also drink a tea of its leaves, which very much refreshes them in violent heats. The tanners boil the roots together with their skins, which very much thicken them: It also yields a black dye. It loves to grow in moist boggy places, and by rivers. The other sort they apply the juice of to ease the pain, and stop the immoderate flux of the piles, taking it inwardly, and outwardly applied as a poultice. The dyers mix it in their compositions to dye black. It grows about a yard high.

PAPAWS.

1. *Papaya major.* They are called trees because they grow as high, but are of no durable substance, and so soft that one stroke of an axe will cut through them. The flowers are of a yellow colour, and ad-

here

here clofe to the body, having no foot-ftalk; then the fruit comes, upon a fhort foot-ftalk, growing in clufters, of a verdant green without-fide; but, when full ripe, they turn yellow, and reddifh on that fide next the fun; it hath a great number of round foft blackifh feeds, about the bignefs of a pepper-corn, lying in a foft pappy fubftance. The outfide peel, cut thin, makes fine green tarts; the inward part makes fauce for pork, fo refembling in colour and tafte apple-fauce, as not to be known to the contrary; it is alfo ufed for goofe or duck. When it is thoroughly ripe, it may be eaten raw, having a pleafant juicy flavour, like fome apples. All thefe trees are very milky; for if you pull off a leaf, there effufe feveral drops of white milk, and the fame when you pull off the fruit. Its milk takes away warts (being very fharp and corrofive), kills ring-worms, and takes off films on the eyes.

2. Spreads itfelf in flowers, and it is very rare to fee any fruit upon them, and thofe fmall and long. The flowers are preferved with fugar, and make a fine fweetmeat.

3. Is the female wild papaw, which is every way like the other female, but only its fruit is much fmaller and rounder, and when ripe is food for birds. They grow wild in the woods.

4. The male wild papaw, which grows like the former.

PARAGUAY TEA.

Since the South-Sea company fet up in England, this herb came to be known there, and was at the time cried up for the beft of teas. I knew a gentleman that fancied, by drinking of Paraguay tea, it broke the ftone he had in his bladder; indeed, I faw him often void

I 4   fmall

fmall fhelly pieces of ftone, that looked as if it fcaled or feparated from the outfide of another; but let the virtues of this plant be what they will, it brings great fums of money to thofe that trade in it at Santa Fé. It is brought thither up the river Plate. There are two forts of it; the one called *yerva-con-Palos*, the other, which is finer and of more virtue, is called *yerva Caamini*. This laft is brought from the lands belonging to the Jefuits. The great confumption of it is between La Paz and Cafco, where it is worth half as much more as the other, which is fent from Potofi to La Paz. There come yearly into Peru from Paraguay, the place where it grows and has its name, above 50,000 arrobas, being 12,000 *cwt.* of both forts; whereof at leaft one third is of that fort called *Caamini*, without reckoning 25,000 arrobas of that of Palos, which goes to Chili. They pay for each parcel, containing fix or feven arrobas, four ryals (which we call in Jamaica bits); being the duty called *alcavala*, or a rate upon goods fold, which, with the charge of carriage, being above 600 leagues, doubles the firft price, making it about two pieces of eight the arroba; fo that at Potofi it comes to about five pieces of eight the arroba. The carriage is commonly by carts, which carry 150 arrobas, from Santa Fé to Xuxui, the laft town of the province of Tucuman; and, from thence to Potofi, 100 leagues farther, it is carried on mules.

Passion-Flowers.

1 The granadilla, fpoken of before.

2. Thofe called pops, becaufe, if you fqueeze the fruit, it pops off, being hollow. The flower hath a fine purple thrum, like a fringe, and a crofs one in the centre of the flower, with a reprefentation of three nails; and therefore hath its name of paffion-flower,

repre-

representing the nails and crofs made ufe of to put our
Saviour to death. There are many different forts of
thefe flowers.

Payco Herba,

Or Indian plantain for the ftone, is a plant of an
indifferent fize, the leaf whereof is very much jagged;
it fmells like a rotten lemon. Its decoction is a fudorific, and very good againft pleurifies; it is alfo excellent for the cholic and ftone. Much of it grows
in Chili.

Peach-Tree.

There is great plenty of thefe trees in North-America. The leaves, decocted, are faid to be a fpecific
for the cholic or belly-ache; fo is alfo the fyrup made
of the flowers, which cleanfes fucking childrens ftomachs
that are apt to puke or throw up their food; it alfo
purges watery humours. I never faw but one peach-
tree in Jamaica, and I never faw or heard of it bearing any fruit.

Pease.

Befides the forts fpoken of amongft the beans, there
are fome that are more properly called peafe. Englifh peafe grow but very indifferently in the fouthern
parts of America; nay, even in Jamaica, they have
nothing in the tafte of the fweetnefs that they have in
England, and therefore they prefer the calavances before them.

Pellitory of the Wall.

American pellitory differs little or nothing from that
in Europe. It hath a fpecific quality to cure the ftrangury and dropfy, expelling gravel or flime from the
reins

teins and bladder; and is alfo good againft coughs, and pains of the pleura, liver, fpleen, and womb. It grows on the fides of fhady rocks.

PENGUINS.

The fruit is good to clean a fore mouth, if it can be endured. A little of the juice, dropped into wa-ter, quenches thirft and heat of fevers; a fpoonful of the juice, with a little fugar, given to children, kills worms, cleanfes and heals the thrufh, or any ulcers of the mouth or throat. They are very diuretic; and the juice, given in rhenifh wine with fugar, brings down the terms in women fo powerfully as to caufe abortion, if given in too great a dofe. Both wine and vinegar might be made from the fruit; and from the leaves might be made a fine flaxen filk, as fine or finer than from the filk-grafs.

PENNYROYAL.

Befides the garden pennyroyal, there are two forts. They refemble it in its leaves, but no way in its biting pungent tafte; and, having flowers like the *amaran-thus*, I take them to partake more of the nature of thofe than of pennyroyal.

PEPPER-GRASS.

This plant is fo called from its hot biting tafte, like pepper; but I think it taftes more like *taragon*, or the land-crefs. Sir H. Sloane makes it to be a *fciatica* crefs. *Sciatica* crefs had its name (as we may fuppofe) from its great efficacy and power againft the hip-gout. It is alfo a great provoker of urine, and cures the fcurvy and dropfy; the juice is excellent in cutane-ous diftempers, mixed with oil of wax. It grows in great plenty fpontaneoufly in moft parts of America:

I faw

I faw a great quantity growing in the church-yard in St. Jago de la Vega.

PEPPERS.

1. *Piper longum arboreum altius folio nervofo minore fpica graciliore et breviore.* This has feveral ftems, rifing twelve or fifteen feet high; they are ftraight, green, fmooth, jointed, and at every joint protuberant or knotty, each joint being about a foot diftant, and being full of a pithy fubftance like elder; fome call it Spanifh elder: Upwards, the joints are at lefs diftance from one another. Towards the top ftand the leaves, one at a joint, upon a fhort foot-ftalk; they are two inches long and one broad, ending in a point; the nerves or fibres of the leaf are very large, running longways, making a pleafant fhow on a very dark-green fmooth leaf, which, when rubbed, is very aromatic. Oppofite to the leaves comes a julus, about an inch long, flender, and of a yellowifh pale colour, refembling long pepper. The leaves and fruit are very hot, and, decocted and drank, are good in the cholic or belly-ache, and in all hydropical difeafes. It alfo makes excellent baths againft all forts of fwellings; it ftrengthens and corroborates the parts.

2. *Piper longum racemofum malvaceum.* This is commonly called *Santa Maria*, from its great virtues. Its leaves are cordated, or more of the fhape of horfes hoofs, foft, of a dark-green colour like the mallow, and refemble the Englifh colt-foot, but much larger, being about feven or eight inches diameter. It loves to grow in fhady places. The leaves, being very foft and large, are applied to the head when it aches, and immediately take away the pain; the fame it doth in the gout: They are thought to eafe pain in every affected part, and therefore are efteemed as a very

rare

rare remedy by all Indians and negroes, as well as planters. If the julus or pepper be fcalded in water, and dried in the fun, they grow ftronger, and more durable for ufe. The root fmells like clover, and is hot to the third degree, and reckoned a counter-poifon. Being of thin and fubtle parts, it opens all obftructions; if bruifed and applied as a poultice to any difeafed part, it ripens and breaks the fwelling, and cleanfes the part. The juice, or an ointment made of the leaves, cures burns, fcalds, or any inflammation. The leaves, in a clyfter, are more emollient than mallows.

3. *Piper longum humilius fructu fummitate caulis. predeunte.* This has a creeping jointed root; the ftalks are round and green, jointed, rifing feldom above a foot high; the leaves are thick, fucculent, fmooth, and of a dark-green colour, having fome vifible veins on the upper furface like thofe of the water-plantain, and fometimes notched at the upper end of the leaf. At the top of the ftalk comes out a flender four-inch fpike julus, or *ligula*, like thofe of *ophioglofium*, or fome of the long peppers, of a fweet fmell, and fharp to the tafte like them, and withal fomewhat balfamic; the plant rubbed fmells very gratefully. It is hot in the fourth degree, and dry in the third. It ftrengthens the heart, heats the ftomach, and gives a fweet breath; attenuates grofs and thick humours; refifts poifon, the iliac-paffion, and cholic; is diuretic; helps the *cutame-nia* or menfes in women, helps birth, expels the dead child, opens obftructions, and cures pains from cold; it takes away the cold fit of an ague.

See Capficum Peppers.

PEUMO.

In Chili is a tree called Peumo; it bears a red fruit in the fhape of an olive. A decoction of the bark cures

the

the dropfy ; the timber of it is ufed for building of fhips.

Physic-Nuts.

Some call them tyle-berries of India. They purge ftrongly upwards and downwards, given from three to five; they may be candied over, and given unknown to nice palates ; if the inward film be taken out, they will work more gently. The beft way of preparing them is, firft to torrify them ; then take off the outward fkin and inward film, that is, the fprout or *punctum faliens*; then bruife them in a mortar, and fteep them in Madeira wine; and they will purge well all grofs humours. They afford great quantities of oil, which may be got by boiling or expreffion, and which purges ftrongly; this oil they ufe or burn in their lamps in Brafil. If you rub the ftomach with the oil, it will purge and kill worms ; it cures the itch, and deterges ulcers. There are three or four forts of thefe trees ; but one, in particular, differs very much from the reft, whofe leaves are more divided, and have a very beautiful fcarlet flower : Thefe never grow fo high as the other forts; they are called French phyfic-nuts, and their purging quality is more ftrong than any of the other forts.

Piemento.

It is alfo called Jamaica pepper, or allfpice. It is fo well known, that it is needlefs to give a particular defcription of it. The fruit is excellent againft the cholic, and all cold and undigefted humours of the ftomach and bowels. A decoction of the leaves, or a bath made of them, is good in all old aches and pains of the bones, and healeth old ulcers.

PIGEON-PEASE.

They are fo called from pigeons greedily feeding upon them, but they fomething refemble a broom-pea. From ftrong fibrous roots fpiings up a ftraight woody ftalk, as big as one's finger, five or fix feet high, like the common broom-ftalk, and it hath yellow flow-ers like broom; it hath a yellowifh green pod, about the length and bignefs of Englifh peafe-cods, and its pea is much of the fame bignefs, but flatter or com-preffed on both fides. Their leaves are very thin and foft, of a dark-green, fmelling fomething like a rofe when rubbed; they are about two inches broad in the middle, and about three inches long, coming off ta-pering. They have bloffoms, green peafe, and dry, upon them all at the fame time, and will keep bearing fo for many years, which makes fome call them feven-years peafe; they are very wholefome food. In fhell-ing of them, there is a clammy or gummy fubftance that comes off and fticks to the fingers, hard to be wafhed off. The juice of the leaves, or diftilled water from them, makes an excellent eye-water.

There are alfo two forts of heart-peafe:

1. Sir H. Sloane calls it *pifum decimum, five vefi-carium fructu nigro alba macula notato.* This has a woody, cornered, rough ftalk, taking hold of any tree or fhrub it comes near with its clavicles, and mount-ing to eight or nine feet; the tops then falling down, cover the tree or fhrub it climbeth upon. At about every three inches diftance, it puts forth leaves, cla-vicles, and flowers, at the fame place. The leaves ftand on two and an half inch long foot-ftalks; they are very much divided or laciniated, cut always into nine fections, ftanding three together on the fame common *petiolus,* coming from the end of the foot-
ftalk;

ſtalk; that diviſion of the three oppoſite to the end of the petiolus, or in the middle, is the biggeſt, being two inches long, and one broad where broadeſt, deeply notched or cut in on the edges, of a dark-green colour, very ſmooth, ſoft, and thin; the other two at the baſe being of the ſame ſhape, and only ſmaller. The clavicles ſtand oppoſite to the leaf, being five inches long. *Ex alis foliorum* come the flowers, ſeveral together, ſtanding on three-inch long foot-ſtalks, being white, pentapetalous, and very open. After the flowers follow three-cornered oblong bladders, having in each of them three diſtinct cells; and in every one of theſe lies, faſtened to a membrane, a round dark-brown or black ſeed, about the bigneſs of a ſmall field-pea, having three triangular lines meeting at the centre of a clay-coloured or whitiſh triangular or cordated ſpot (and therefore called *piſum cordatum*), which is at the place where it is joined to the bladder or its *hilus*. The ſeeds of this plant cauſe greater ſleep than opium; bruiſed with water and applied, they eaſe the gout, and coldneſs of the joints with ſtiffneſs; the juice of the leaves, with black cummin ſeed, is good for heart-burning; and mixed with ſugar is good for a cough.

2. The other ſort is *piſum cordatum non veſicarium*. This grows like the former, only it hath a larger pea, with a white hilus, eye, or ſpot. The green leaves bruiſed, or their juice, are good for wounds, being a great vulnerary, and cleanſing. The fruit, bruiſed and put into water, intoxicates fiſh.

PILEWORT.

We have a plant named Indian pilewort, which is called by native Indians *guacatane*. It is white, like unto *pol'um montanum*, but without any ſweet ſcent. Monardus ſaith, it grows in great plenty in Hiſpaniola.

It

It is much commended to help or take away the pain, inflammation, and swelling of the piles, and falling out of the *anus*, by fomenting the part with a decoction of the whole plant, and strewing thereon the dried leaves in powder.

PILLERILLA

Is the name that the Spaniards in Peru give to the *palma Chriſti*, or *ricinus Americanus*. They affirm, that the leaf of it, applied to the breaſts of nurſes, brings milk into them, and, applied to their loins, draws it away.

See Oil-Nut.

PILOSELLA

Is a plant which hath a ſcent like wormwood, but grows like mouſe-ear. Theſe ſometimes cover whole fields in South-America and Chili.

PIMPERNELL.

There are two kinds of this plant growing in America.

1. This ſmall repent, or creeping plant, has round, ſmooth, green, juicy ſtalks, which, at every joint, ſtrike into the earth ſmall white hairy fibres, whereby it draws its nouriſhment, and likewiſe ſmall green ſucculent or juicy leaves, almoſt like thoſe of water-purſlane, being roundiſh, thick, green, ſmooth, and very ſmall, without foot-ſtalks, ſtanding oppoſite to one another towards the end of its ſmall twig. *Ex alis foliorum* come out half-inch foot-ſtalks; and on them, in a calyx conſiſting of two green leaves, a pentapetalous or five-leaved flower, of a pale blue colour, having ſome whitiſh ſtamina within. After this follow a great number of very ſmall flat brown ſeeds, incloſed in a hard brown capſula or caſe, covered by ſome, firſt green,

green, afterwards brown, leaves, which are the peri-anthium or calyx of the flower.

2. Has a very deep-blackifh coloured root, which fends up a round brownifh woody ftem, rifing three or four feet high, being divided into branches on every hand. The leaves come out feveral together, fome greater, fome fmaller, at half an inch diftance, on half-inch long foot-ftalks; they are half an inch long, and a quarter broad at the bafe (where broadeft), of a grafs-green colour, indented about the edges like ger-mander, but fmooth. Oppofite to the leaves come yellow flowers, being ftamineous; after which follows a two-inch long dark pod, or feed-veffel, fhutting like thofe of the fefamum, but more like the fpirit-weed, only having two round fides, and a partition in the middle; in which are two rows of feeds, black and quadrangular. The pod, when ripe, opens at the end, and fcatters the feed like as the fpirit-weed.

Pimpernells are accounted a peculiar remedy againft the plague, and all malignant or peftilential fevers; alfo good againft the bitings of ferpents, efpecially the rattle-fnake, and an excellent wound-herb, ftopping fluxes of humours. Dr. Bowles fays, they cure can-cers; Morrifon fays, they cure phthificks; Quercetan affirms, they ftop immoderate menfes; and Hermius, that they cure madnefs.

PINDALES.

The firft I ever faw of thefe growing was in a ne-gro's plantation, who affirmed, that they grew in great plenty in their country; and they now grow very well in Jamaica. Some call them *gub-a-gubs*; and others ground-nuts, becaufe the nut of them, or fruit that is to be eaten, grows in the ground: Thefe are of the bignefs, colour, and fhape, of a filbert; they are co-

K   vered

vered over in the ground with a thin ciftus or fkin, which contains two or three of them, and many of the ciftufes, with their nuts or kernels, are to be found growing to the roots of one plant. When they are ripe and fit to dig up, the ciftus that contains them is dry, like a withered leaf, which you take off, and then have a kernel, reddifh without-fide and very white within, tafting like an almond, and accounted by fome as good as a piftachio; they are very nourifhing, and accounted provocatives. Some fay, if eaten much, they caufe the head-ache; but I never knew any fuch effect, even by thofe who chiefly lived upon them; for mafters of fhips often feed negroes with them all their voyage; and I have very often eat of them plentifully, and with pleafure, and never found that effect. They may be eaten raw, roafted, or boiled. The oil drawn from them by expreffion is as good as oil of almonds; and the nut, beaten and applied as a poultice, takes away the fling of fcorpions, wafps, or bees.

PINE-APPLE.

A moft delicious fruit, called *ananas.*

PINKS.

We have in America pinks, carnations, and gilly-flowers, growing in gardens; befides which, we have a moft beautiful pink that grows wild in the woods, mixed with white, red, and other colours, in a moft wonderful manner.

PLANTAIN.

The common Englifh plantain grows fpontaneoufly here very well; befides which, we have feveral other forts.

1. *Plantago aquatica,* or water-plantain. It is fo
well

well known in America, that there needs no particular
defcription of it; it grows like thofe in England. You
may find it growing along the river-fides, and in watery places. It is thought to have the fame virtues
with land-plantain; the feed is aftringent, and the
leaves good againft burns, and proper to be applied to
hydropic legs. The juice, applied to breafts, is a great
fecret in drying up and clearing them of milk.

- There is another fort, which Margraave calls,

2. *Planta innominata;* and fome would have this
to be a *phalangium,* or fpider-wort.

3. *Plantago aquatca folio anomalo flore ftipitato
purpureo femine pulverulento.* This has feveral large
white roots, two or three inches long, from which come
feveral leaves, four or five inches long, green, fucculent, and ribbed like plantain-leaves In the centre
of thefe leaves rifes a purple jointed ftalk, a foot and a
half high, having a fpike of purple or carnation flowers three inches long, and at the top three purple petala or leaves; under which is a little fwelling, of a
brown membranaceous fkin or hufk, containing a fine
dufty feed, which is fcattered with the wind.

All thefe plantains are cooling and reftringent, and
therefore good in aneurifms, and falling-out of the
fundament; they ftop fluxes of all forts, and prevent
abortions. The feeds, bruifed and infufed in claret or
Madeira wine, or the juice taken inwardly, and applied
outwardly, abates inflammations.

PLANTAIN-TREE.

This, as well as the banana-tree, hath the name of
mufa, and they are fo alike, that, unlefs perfons are
well acquainted with them, they would not know one
from the other at fight; but the fruit differs, they being
much longer and larger than the banana. The fruit

K 2   of

of this tree is the beft of all Indian food for negroes, and makes them the moft able to perform their labour, and therefore confequently muft be of great nourifh-ment. Roafted before they are ripe, they eat like bread; they are eaten boiled or roafted, and one roaft-ed that is ripe, and buttered, eats very delicious.

If you thruft a knife into the body of one of thefe trees, there will come out a great quantity of clear water, which is very rough and reftringent, ftopping all fort of fluxes: I have advifed perfons fubject to fpit blood to drink frequently of this water, which cured them.

There is a wild fort of thefe trees, but much fmaller, although the leaf is broader than either this or the ba-nana; but they bear no fruit, and therefore are of no value.

PLUM-TREES.

Of which there are feveral forts, but none to com-pare in goodnefs to thofe in Europe.

1. The Spanifh yellow plum.

2. The common deep-red or purple-coloured plum, which comes before any leaves upon the tree. Some of them have a knob at the end, and are called the top-knot plum.

3. Called the hog-plum tree, and is a larger tree than any of the reft, having a large yellow plum, which hath a rankifh fmell, but a pleafant tart tafte. The hogs feeding upon them, they are called hog-plums; fheep alfo feed upon them, when fallen upon the ground. In the year 1716, after a fevere fever had left me, a violent inflammation, pain, and fwelling, feized both my legs, with pitting like the dropfy: I ufed feveral things, to no effect. A negro going through the houfe when I was bathing them, faid, "Mafter, I can cure you,"

you," which I defired he would; and immediately he brought me bark of this tree, with fome of the leaves, and bid me bathe with that. I then made a bath of them, which made the water as red as claret, and very rough in tafte: I kept my legs immerged in the bath as long as I could, covering them with a blanket, and then laid myfelf upon a couch, and had them rubbed very well with warm napkins; I then covered them warm, and fweated very much: I foon found eafe, and fell afleep. In five or fix times repeating this method, I was perfectly recovered, and had the full ftrength and ufe of my legs as well as ever; giving God thanks for his providential care, in beftowing fuch virtues to mean and common plants, and that the knowledge of them fhould be made known to fo vile and mean objects as negro flaves and Indians.

4. Maiden plum.
5. Coco plum.

POISON-BERRIES.

Sir H. Sloane tribes thefe among the jeffamin-trees.

POLYPODIUM

Are of the fern kind, and therefore tribed amongft them. They grow exactly as thofe in England, although they have not oaks to grow upon; I have feen them grow at the bottom of palm-trees, but yet have the fame virtues as thofe in England, which are accounted fpecifics, purgers of melancholy humours and tough phlegm; they open obftructions of the fpleen, and expel wind. A fyrup made of them is good for coughs, fhortnefs of breath, hoarfenefs, and wheezing of the lungs.

POMEGRANATES.

Thefe grow in great plenty with us, and as good as

in any part of the world; they have a large scarlet flower, and are restringent.

POND or RIVER WEED.

These weeds grow in most rivers in America. They are cooling and drying, stop fluxes, and, outwardly applied, take away all inflammations of the skin, &c.

POPES HEADS.

Some call them Turks heads, for they something resemble them when they have their turbans on. They grow close to the ground, beset all round with prickles, and are well known in America, growing on the worst salt sandy ground, where nothing else will grow but prickly pears or *opuntias*. They have on the top a purple flower, like an artichoke or globe-thistle, and a small red or crimson cod or fruit, of the shape of a long red pepper, which hath a very pleasant tart taste, and is very cooling. It is hollow, like the capsicum or long red pepper, and full of small black seeds.

POPONAX.

This is a name, but very erroneous, that they in Jamaica give to a plant which is of the *acacia* kind, and is more exactly like the Egyptian *acacia*, or thorn. It is reported, that a certain person brought the seed of it to Jamaica, and planted it, and said, if he lived to see it grow, he should get an estate by it; but how, remains a mystery to this day, unless it is for its dying quality; its flowers are indeed very odoriferous. The dyers use the husk of the pods to dye black; they also soak some of the pods all night in water, then mix a little alum with it, and boil it to a due thickness; which makes a very fine black and strong ink. I have often made it, and wrote with it, and observed it never fades

or

or turns yellow, as copperas ink will. I carried some of the pods with me to England in 17 7, and gave them to a dyer, who tried them, and said, they exceeded galls for dying of linen, and, if they wou'd come as cheap, would be preferable: But I alfo obferved, the worms deftroyed the pods and feeds quickly.

It is certain that the *fuccus acacia*, that is one of the ingredients of mithridate, and Venice treacle, is only the hardened juice extracted by decoction of the *acacia* or Egyptian thorn, which I take to be this tree, or at leaft to be as good, if not better, having rather a more reftringent quality, and therefore proper in all forts of fluxes.

The name poponax, that they give to this plant, I take to be the corrupted word of *opoponax*, which is a gum, or infpiffated juice, of a plant called *panax heraclium*; but this is not the tree.

There is another fort called *acacia*, but more reprefents a wild tamarind, and therefore the planters in Jamaica call it fo; for the fruit is a longifh pod, which, when ripe, opens and turns infide out; it is of a glorious red colour. There is alfo another fort, very improperly called wild tamarind, which is a certain *acacia*, with very large prickles; but I think the flower of this tree is not fo fweet-fcented as that they called poponax. Thefe are fine large fpreading trees, as big as the Englifh elms, but much more fhady and fpreading. Both the bark and roots of this tree ftink worfe than *affa fœtida*; they are of a reddifh colour, and dye red. The wood is good timber.

Poppy.

We have a plant that grows like the Englifh common prickly thiftle, but its flower is yellow, in the fhape of the field poppy; and after the flower come

heads

heads that are as big as a walnut, armed thick with sharper prickles than the *stramonium*. Its seeds are like the black poppy, but much more narcotic.

The whole plant is milky, but of a yellow colour; which, mixed with womens milk, and dropped into the eye, clears the sight, and takes off spots or films: It may be for this reason it is called *argemone*. It also wastes fungusses, or proud flesh. The distilled water, with the tops of wild tamarinds, makes a good eye-water.

The fruit or head is called *figo del inferno*, or *ficus infernalis*, and well it may, for it contains seeds enough to send any that should take them wilfully to *inferno*, being much stronger than any opium, as was lately discovered in Jamaica in the following manner: A negro man, who had run away some time from his master, lived by stealing of stock; one night he came to a sheep-pen, where there was only a poor old negro man to look after it, to whom he said, he must have a sheep to-night; the old man not being able to resist him, gave him good words, and asked him to smoak a pipe, which he filled for him, putting in a quantity of the seeds of this plant, and before he had smoaked out his pipe, he fell into a sound sleep, not easily to be awakened; upon which, as the old fellow knew very well the effect, he ran to a neighbouring pen, and getting ropes and assistance, they secured him before he was thoroughly awake; and when he was, he cursed and swore, saying the old fellow was an *obeah* man, and had bewitched him. I saw a fat steer drop down dead of a sudden, fetching two or three staggers, foamed at the mouth, and died immediately: I ordered them to cut his throat; and, after opening him, in his stomach were found several handsful of the seed of this plant, which I supposed killed him.

POQUETT.

Poquett

Is a fort of gold-button, or female foothernwood, with green checquered leaves, which dyes yellow, and holds well.   The ftone of it dyes green.

Potatoes, or Batatas.

Potatoes grow in great plenty in moft parts of America, and are a convolvulus plant, with a bell flower; but as they put nothing for them to run upon, they creep and fpread upon the ground, covering it fo, that it deftroys grafs that would grow there.   They are of feveral colours; the roots are fome red, fome very white, and fome yellowifh, or fulphur colour; they exceed, in my opinion, the Irifh or Englifh *batata.* They are one of the chief bread kind, as they call it, in America, and are food for white and black; they are very fine when baked.   The flips or vines they feed hogs and rabbits with; and an excellent drink is made of the roots, called *mobby.*

Prickly White Wood.

This grows like the prickly yellow wood, only the wood within-fide is very white.   It hath fmall bunchy flowers; after which come bunches of black triangular feeds, in fhape and bignefs of buck-wheat; they are hotter upon the tongue than any Guinea-pepper, and negroes take them for the cholic.   The roots of the prickly woods are ufed in venereal cafes.

Prickly Withe,

Which fome call prickly pear withe.   In the centre of the green fucculent part there is a ftrong wire withe, which planters ufe, and is very lafting.

Prickly

PRICKLY WOOD.

There are feveral trees in America that go by the name of prickly woods; but the moft common fort, and what is moftly known, is called

PRICKLY YELLOW WOOD.

It hath a leaf like Englifh afh; the outfide bark is brownifh, fet full of protuberances, about an inch or two inches long, and as thick as a man's finger; at the end of which is a fhort fharp prickle. The infide wood is very yellow.

PUMKIN.

We have pumkins of various fizes and fhapes, as large as any in England, and as good. This fruit is much eaten; but too much is apt to furfeit, and to caufe fevers.

PURSLANE.

This plant, which is fo much taken care of in England to cultivate in their gardens, grows wild in moft parts of South America. It is a cooling and moiftening herb, therefore good in burning fevers. I often prefcribed, in America, the diftilled water in fevers, efpecially where a flux attended them. It takes away the ftrangury, as well as the heat and fcalding of urine in ardent fevers. Eaten raw, it cures teeth that are fet on edge, and faftens them. The juice of the herb is fingularly good in inflammations and venereal ulcers. The herb, bruifed and applied to the forehead and temples, allays the exceffive heat and pains that occafion want of reft and fleep, and, applied to the eyes, takes away rednefs and inflammations. The juice, mixed with vinegar, takes away the St. Anthony's

fire,

fire, and pimples in the face. The juice, with the oil of roses, takes out the fire of burnings by gunpowder, lightning, or scalding, but if it were mixed with goose-greafe it would do better; the juice alfo, made up into pills with gum tragacanth and arabic, cures thofe that evacuate or spit blood. The feed is more effectual than the herb, and is of fingular ufe for all the purpofes above mentioned.

Quamoclit.

This is a convolvulus plant. It rifes firft with two oblong broadifh leaves conjoined, refembling the fruit of the maple, which remain long without fading, even after the plant begins to wind itfelf round its prop. The other leaves fhoot from the purplifh viny ftalks, in an alternate order; they are winged, finely cut and divided, of a dark-green colour, but the young leaves are yellowifh, or pale-coloured, having at firft but few divifions or wings; afterwards, they are fplit into feveral, to the number of thirteen, with one at the top; the firft divifions are ufually forked. The flowers are of a moft elegant beautiful red, fhooting alternately from the joints of the viny ftalks, fometimes fingle, fometimes two together, monopetalous or bell-flowered, all in one leaf, fhaped like a funnel, and divided into feveral fegments. From the flower-cup the pointal rifes, which is fixed like a nail in the bottom part of the flower, and has five yellowifh threads and chives within. They are fucceeded by an oblong fruit, ftanding in a fcaly cup, with a tough bark or fkin like the other bindweeds, which inclofes four oblong black and hard feeds. The tafte of the herb itfelf is fweetifh and moderately nitrous: The whole plant fwells with a thin pale milky juice. The root is a ftrong purge.

QUESNOA,

QUESNOA, or QUINA,

Is a little white feed like that of the muſtard, but not ſo ſmooth; which is good againſt falls and bruiſes, and the ſpaſms, a ſort of convulſions.

QUILLAY.

This is a tree, the leaf whereof ſomewhat reſembles that of the ever-green oak. Its bark ferments and lathers like ſoap, and is better for waſhing woollen clothes, but not for linen, which it makes yellow. All the Indians make uſe of it for waſhing their hair and cleanſing their heads, and it is thought to be that which makes their heads ſo black.

QUINCHAMALI.

This is a ſort of ſantolina, or dwarf-cypreſs, bearing a yellow and red flower. The virtue of this plant is, that if any man happen to have a violent fall, which occaſions him to bleed at the noſe, or inwardly, the decoction of this herb, drank plentifully, is an infallible remedy.

QUINQUINA.

This is what is commonly called jeſuits bark, or Peruvian bark. It is the outward bark of a tree that grows in Peru, and chiefly in the province of Quito, upon the mountains near the city of Loxa, and was firſt brought into Europe by cardinal Lago, a Jeſuit, in the year 1650. The tree is about the ſize of European cherry-trees, the leaves round and indented, and it hath a long reddiſh flower, from whence ariſes a kind of pod or fruit, in which is found a whitiſh kernel, like an almond, flat, with a thin ſkin. Chooſe that which is a lively-bright cinnamon colour within-ſide, and darkiſh without, but, which is called quill-bark, and comes from the branches of the tree; fee that it be heavy and found, dry and firm, breaks a little fhining, and hath a little white fpeck like mofs, or fome fmall fern threads flicking to the outfide bark or fkin, and is very bitter in tafte, with an aftringent rough ftipticity upon the tongue: Refufe that which is full of chives when broke, of a dark or ruffet colour, thick, flat, and very heavy.

There is another fort of this bark, which comes from the mountains of Potofi: It is much browner and thinner than the former, more bitter and aromatic, and much more fcarce and difficult to be got. This is much ftronger in operation than the other; one ounce will do as much as three ounces of the common fort. The firft time I faw it was in a galleon, that lay in Port-Royal harbour in Jamaica, in the year 1709, taken by admiral Wager.

RAGWORT.

This is alfo called St. James's wort, and there are many forts of them. They are good wound-herbs, are much commended in quinfies, ulcerated mouths and throats, and difcufs hard fwellings.

There is a ragwort grows in Chili, whofe flowers are yellow: The Indians call it *nillque*, and make a tea of it, which they drink after the cold fit of an ague, and it abates the heat that follows. It grows on the rugged fea-banks of Chili.

RAMOON.

This is a name they give to a tree that grows in Jamaica, well known to the planters, who give the tops and branches of it to their cattle, which makes them fat. The medicinal quality as yet is not known; but I hope in time fome curious perfon will make

fome

some strict enquiry into it, and make some experiments on it.

RAMPIONS.

There is great variety of these plants in America, but of very little medical use. Rampions have the leaves of throat-wort, and purplish flowers. The distilled water of the roots, leaves, and flowers, of these plants increases milk in women; a decoction of the whole plant is cooling and abstersive, and therefore good against inflammations, sores, and ulcers of the mouth and throat.

RAQUETTE.

This is one of the dildo trees, and that which Sir H. Sloane calls *cereus crasissimus, &c.* and which Piso calls *Jamacara;* but Herman calls it *cereus erectus fructu rubro non spinoso;* therefore the fruit of this cannot be the *higas de Tuna,* or Tuna figs, as they call them in America; for they are full of prickles, and therefore are those that we call in Jamaica, prickly pears; but this is supposed to be the plant that gum Euphorbium is got from. Euphorbium is so called from Euphorbius, physician to King Juba, who first introduced it into practice and use: it was this physician that cured Augustus Cæsar of a distemper. Choose that which is white, bright, and clear; that also which is of a yellowish colour is good, if it be so sharp that, upon a small touch upon the tongue, it burns and heats it; the older the better.

REEDS.

We have several sorts, which are most exactly like those in England, and, having the same virtues and uses, I therefore refer to those who have written of them at large.

REILBON

Reilbon

Is a fort of madder that grows in Peru; the leaf of it is fmaller than ours. They ufe it, as the dyers in England, to dye red.

Rest-Harrow.

We have a fort of this plant that differs much from thofe in England, having no prickles. Thefe plants are clammy, and fmell like the ordinary *ciflus*. They have a peculiar quality to provoke urine, to diffolve vifcofities and tartarous humours in the reins and urinary paffages, and to open obftructions. Ray affirms, that it cures *hernia carnofa*.

Rice.

Rice grows as well in America as it doth in Africa and other parts. About twenty years paft, I fowed fome in a moift parcel of ground in Jamaica; but happening to plant out of time, it grew very rank, and did not bear. I cut it down clofe to the ground, and gave it to my horfes, who eat it as well as Guinea-corn blades. Afterwards it grew up, and, at the ufual or proper time, it bore an extraordinary quantity of grain, which was bearded like barley, which that with its outward hufk is taken off, and then it is quite white. The Spaniards and Portuguefe call it *arroz*, of which they make a fpirit called arrack; the Arabians call it *arz*, and *arzi*. It is cooling and reftringent; an emulfion made of it is good againft the ftrangury from cantharides; the fine meal or flour takes away the marks of the fmall-pox.

Ricinus.

There are many kinds or forts of *ricinus* in America.

1. That

1. That commonly called oil-nut-tree, which has been already described.

2. *Ricinus Americanus major caule virescente.* This differs only, that the stalks of this are very green and the other reddish, and the fruit rather less.

3. The physic-nut, described before.

4. This differs very little from the former, only the leaf is thinner, and more divided at the ends, like briony leaves, and has a fine scarlet-coloured flower. The fruit is an easier purge than the common physic-nut; the flowers, dried and powdered, purge hydropic water plentifully.

5. The wild caffada, described before. This plant resembles the slaves-acre, that grows in Provence and Languedoc; but that has six or seven points, when old or full grown, and this but five.

6. The true caffada.

7. Wild rosemary.

ROCKET.

I never could find out but one sort of rocket in America, and that very little notice taken of it, being a sort of sea-rocket: It grows like that in the Mediterranean sea, and is something like the *eruca marina Anglica.* They grow in salt ground near the sea. They purge strongly; the distilled water, four ounces drank warm, takes away the cholic, provokes urine, and kills worms.

ROSEMARY.

Besides the garden rosemary, we have a wild Spanish rosemary. This shrub grows as big as one's arm, covered with a light-brown smooth bark, rising five or six feet high, having many white branches, beset with leaves about an inch distance from one another; they

are

are two inches long, and a quarter of an inch wide, exactly like rofemary, but very white underneath, and green on the top or upper-fide as rof mary, and ftanding upon the ftalk as they do; the tops of the branches, for three inches length, are fet thick with fmall white flowers, made up of many ftamina; the flower is five-leaved. After this follows a tricoccous fruit, very fmall, fticking clofe to the ftalk, fmooth and whitifh, each of the three fides containing an oblong brown fhining feed: The whole plant fmells very gratefully and ftrong. It is ufed very much in all forts of medicated baths and fomentations for hydropical perfons: the powder of the dried leaves is a fpecific in the cholic, and in all cold watery undigefted humours, having all the virtues of rofemary.

The fecond fort has feveral fmall woody branches, about four or five feet long, fometimes rifing upright, and fometimes lying along the furface of the earth, having a grey bark; the twigs have leaves at their ends, about an inch and half long and an inch broad, which makes them oval, fnipt about the edges, and of a very dark green, fomething like tree-germander. The flowers confift of fix green ftamina, coming from the fame centre, ftanding in a pentaphyllous calyx, coming out *ex alis foliorum* by very fhort foot-ftalks; to which follows a green tricoccous feed, which afterwards grows as big as that of *heliotropium tricoccum,* only it is fmooth, and of a very pleafant pale-purple colour. The leaves of this plant, bruifed, are very odoriferous. This much refembles the *teucrium,* or tree-germander, and has much the fame virtue, but is rather hotter.

ROUNCEVALS

Are a fort of peafe, growing in America, in fhape of the Englifh rouncevals; but the pod differs, and is like the calavances.

L   RUE.

Rue.

Besides the common garden rue, which grows very well with us, there are many wild rues, that grow in great plenty upon rocks in the mountains in America, which are commonly called wall-rues, and are tribed among the fern kind. Sir H. Sloane takes notice of four or five forts growing in Jamaica. These wall or wild rues are accounted fpecifics againft poifons, whether inwardly taken or outwardly received, by the bitings of ferpents or other venomous creatures. The following electuary is admirable for the fame purpofes : *Powder of thefe rues, four ounces; zedoary, contra-yerva of Jamaica, Virginia fnake-root, and Indian arrow-root, of each, in fine powder, one ounce; faffron, in powder, half an ounce; cochineal, a quarter of an ounce; the rob or juice of thefe, with fugar or honey, make an electuary according to art; the dofe is from one drachm to two, or as much as will lie upon the point of a broad knife, drinking a glafs of Madeira wine after it.* This electuary is excellent againft the plague or any peftilential fever, drives out the fmall-pox or meafles, fortifies the heart and refrefhes the fpirits, opens obftructions, cures the jaundice and cholic, and takes away hyfteric fits.

Rupture-Wort.

There are few or none of thefe plants to be found in America. The only one is taken notice of by Sir H. Sloane; it is a water rupture-wort, growing on the banks of moft rivers and wet places. The roots of thefe are many, fmall, and hairy; the ftalks green, round, erect, lucid or almoft tranfparent, about a foot high, having on each fide, alternately, a fmall branch, and oppofite to it a tuft of leaves; and out of the branches, after the fame manner, come twigs, having very fmall, green, lucid leaves, like thofe of *polygonum,* or knot-grafs, only fmaller in every part, very thick fet one againft another. The flowers come out, *ex foliorum alis,* on very fmall petioli, either red-difh or green, and tetrapetalous, but fo fmall as can hardly be difcerned; the feed follows, as fmall as duft. This plant is very aftringent to the tafte.

RUSHES,

Of which we have feveral forts, as you may fee in Sir Hans Sloane's Natural Hiftory of Jamaica, *p.* 121 and 122.

1. The *apoyomatlis,* or *phatzifiranda* of Hernandez. It hath a red knobby root, which hath a very odoriferous fmell, exceeding *calamus aromaticus,* and hath the fame virtues; but I think it fmells like Florence *orice.* The ftalk is like Englifh common rufhes. This is a great antidote againft poifon, expels wind, takes away the cholic, and fortifies the ftomach, caufing a good digeftion.

2. That which the negroes call *adru.*

3. The rufh with which negroes commonly bottom chairs, and make mats, in Jamaica.

4. Which is a cat's tail, or reed-mace. Thefe latter rufhes are very aftringent, and the feeds ftupifying; mixed with butter, or any other proper thing, they kill mice; mixed with hog's fat, but better with goofe-greafe, they take away burning and fcaldings.

SAFFRON.

That which grows in America comes far fhort in goodnefs to that in England. Here alfo grow in great plenty the *cnicus,* five *carthamus fativus,* and *cnicus perennis.* The flowers of *carthamus* are much ufed by

the

the Spaniards (who call them baftard faffron) in all
their broths, to give them a yellow colour, which they
do; they are alfo ufed for dying.   The feed is what is
chiefly ufed in phyfic, or rather the kernel within the
feed, which, beaten into an emulfion with honeyed
water, or with the broth of a pullet, and taken fafting, opens the body, and purges watery and phlegma-
tic humours, both upwards and downwards; the feeds
do the fame clyfterways; an electuary or lohoch, made
with fugar or honey, and almonds and pine-kernels,
cleanfes the breaft and lungs of phlegm; a drachm of
the dried flowers taken, cures the jaundice; the con-
fect, called *diacarthamum*, is a very great medicine to
purge choler and phlegm, as alfo watery humours.
Parrots delight to feed upon them.

SAGE.

English garden fage grows but very indifferently in
the fouthern parts of America, and much care muft
be taken of it to make it grow; but we have feveral
fhrubs called wild fages, their leaves being much like
garden fage, but more odoriferous.

1. The firft is a fhrub, full of branches, growing to
five or fix feet high, and fet full of leaves, very rough
and jagged as a nettle, but in fhape of fage; at the
top of the branches come out many yellow or golden
flowers, confifting of many leaves; after which come
clufters of fmall greenifh berries, like honeyfuckles or
woodbines; they are black when ripe, containing
fmall feeds.   For its great qualities it may well be
called a fage, having all its virtues. It makes an excellent tea to ftrengthen the ftomach; outwardly, if you
apply the bruifed herb like a poultice, it will cleanfe
the worft of ulcers, and heal any wound. The decoction is an excellent bath to ftrengthen the limbs.

2. A

2. A large wild fage, with white flowers, and commonly called in Jamaica *jack in the bufh*.

St. John's Wort.

I have feen a flender plant, which could hardly fupport itfelf, growing amongft bufhes, which had a flower exactly like St. John's wort, but its fruit was like fycamore.

Saloman's Seal.

This plant is well known to the negroes in Jamaica, who eat it boiled.

Sampier.

There is nothing more common in America than fampiers of feveral forts, which grow in all the falt grounds by the fea; but the chiefeft is the common fea fampier, the fame that grows in England; and I have eat of it pickled in Jamaica, as good as any in Europe. It hath the fame virtues.

There is another fort, which refembles the Englifh *kali, kelp,* or glafs-wort; another fort hath a thick juicy faltifh leaf, in fhape of purflane, and is good pickled; another fort hath a turnfole leaf. Sampiers help ftoppage of urine, *&c.*

Sargassa, *or* Zargasso.

This is a fea-weed, of which I took up much in going from Jamaica through the Gulph. It is ufually about a foot high, having tough, fmall, dark-brown or blackifh ftalks, on which are feveral fmall leaves, ferrated about the edges, of a dark-brown colour. It has many round air-bladders coming out from the ftalk, on fmall foot-ftalks, very much like to lentils, which gave it the name. The whole herb, when dry, is hard and brittle. A feaman affirmed to me, that, by eating of it, he was cured of a ftoppage of urine, and brought away a great deal of fand and grofs hu-mours.

Sarsaparilla.

This plant is commonly known by this name, but fome call it *fmilax*, it being thought to be of the fpe-cies of the China-root. The ftalk is long, ferpentine, woody, and prickly, climbing like a vine or a con-volvulus upon every fhrub or tree it is near; the flowers are white, and produce a berry, round and flefhy, like fmall cherries, green at firft, and as they ripen turn a little reddifh, and when full ripe are black, containing one or two ftony feeds, of a whitifh-yel-low, having a white kernel. Although this plant grows in great plenty in Brafil, and other parts of America, yet it is not much taken notice of by the native Indians, the ufe of it having been found out and improved by the expert phyficians of Portugal and Spain. There are two fpecies of it; the ftalks are alike, but different in bignefs and fhape of the leaf. The beft is that of Honduras, which hath a ftalk whofe outfide is very prickly, creeping on the banks in fhady woody places; the leaves are cordated, and of a different length and breadth, of a frefh green on the upper fide, the under fide more pale, growing fingle on the ftalks, alternatively, at a good diftance from one another, having large ribs in fhape and man-ner of *malabathrum*, or Indian leaf; at the foot-ftalk of each leaf grow two fmall long tendrils or clavicles, by which it holds faft to the plant it joins to. The flowers grow in bunches, and are whitifh; from thence follow the berries in bunches, firft green, then red, and at laft black, round, and wrinkled or fhrivelled

like

like dry cherries, containing one or two hard stones, of a whitish-yellow colour, with a hard white kernel, like a small almond. The root of this plant is what is made use of, and it is long and smooth, when first gathered, like a withe, without any prickles, having a thin skin or bark; between that, and a small wire withe in the middle, lies a white mealy substance when dry, which is all that is of use; and of this, ptisans or diet-drinks are made, to sweeten the blood, and for curing venereal diseases. The powder of the root is given, from a drachm to two, to cause sweat. It is reckoned a great alkali, to correct all saline pungent salts in the fluids of the body, and by that means cures venereal diseases, helps rheumatism, catarrhs, gouts, and all diseases proceeding from a superabounding saline acid in the blood and juices of the body.

SASSAFRAS.

Some call this ague-tree, from the Indians performing great cures in agues and swelled legs with a decoction of the bark and root of it. The whole plant is a great anti-venereal and antiscorbutic, opening all obstructions, especially the distilled spirit and oil. I remember that my father cured many scorbutic people with a very strong decoction of the root of the tree, some that were so crippled with pain and swellings in their knees, that they were forced to use crutches; it also cures a *paraplegia,* or numb-palsy.

SAVANNA-FLOWER.

This is so called in Jamaica, because it is found all the year round in blossom, in open savannas. It is too well known, and it is pity that ever the negro or Indian slaves should know it, being so rank a poison: I saw two drachms of the expressed juice given to

a dog,

a dog, which killed him in eight minutes time; but it may be so given, that it shall not destroy a person in many days, weeks, months, or years. Some years past, a practitioner of physic was poisoned with this plant by his negro woman, who had so ordered it that it did not dispatch him quickly, but he was seized with violent gripings, inclining to vomit, and loss of appetite; afterwards, he had small convulsions in several parts of him, a hectic fever, and continual wasting of his flesh. Knowing that I had made it my business some years to find out the virtues of plants, especially antidotes, he sent to me for advice; upon which I sent him some *nhandiroba* kernels to infuse in wine, and drink frequently of, which cured him in time; but it was a considerable while before his convulsive fits left him. The whole plant is full of milk; it is always green, and no creature will meddle with it.

SCABIOUS.

We have a sort of *scabious* grows in Jamaica, that has a round, striated, rough, and pretty large stem, rising to three feet high, having, towards the bottom, several leaves, set without order, on a half-inch long stalk. The leaves are five inches long, and two broad to the middle (where broadest), from a narrow beginning increasing to the middle, and then decreasing to the end, indented about the edges, being rough about, having the surface scabrous or corrugated, after the manner of sage or fox-glove, and woolly underneath; towards the top, the leaves are smaller. Out of their *alæ* come hoary stalks, an inch long, supporting a round head of many white tubulous oblong flowers, each flower standing in a chaffy calyx or perianthium, made up of several dry brownish membranes, which afterwards contains three or four small, oblong, smooth,

and

and shining grey seeds, having a few pappous hairs on their upper ends. This *scabious* is almost like the Spanish *scabious*, only the leaves are not so much divided and jagged as the Spanish. Parkinson faith, that *scabious's* variation and difference confisteth chiefly in the leaves and flowers, not much differing in taste the one from the other, and therefore their virtues are to be accounted alike. They are hot and dry, of an opening, cleansing, digesting, and attenuating quality, whereby they are effectual for all forts of coughs and shortness of breath. The following decoction is very good: *Dried scabious, one handful; liquorice-root, sliced, one ounce; figs, twelve; annifeeds and fennel-feed, of each an ounce, bruised; oriceroots, cut in thin slices, half an ounce; let them steep all night in a quart of wine; then boil the next day, until a third part is consumed; decant, and sweeten with honey or fugar; whereof take a draught morning and evening, for the diseases above mentioned.* Clarified juice of *scabious*, four ounces, taken with a drachm of Venice treacle, defends from the infection of the plague or pestilence. The herb also, bruised and applied to any carbuncle or plague swelling, is found by many experiences, faith Parkinson, to dissolve or break it within the space of three hours; the same, taken inwardly or outwardly, expels and takes away the poison of venomous creatures. A decoction of the roots, drank for forty days, cures leprosies, and all breakings-out; the juice does the same, and heals inward bruises.

SCAMMONY.

Scammony is the inspissated or thickened juice of a convolvulus plant. People differ in their opinion of this plant; some affirm, it is got only from one particular plant; others say, there are several plants that scammony is made from; such as follows:

1. Con-

1. *Convolvulus marinus catharticus folio rotundo flore purpureo.* It grows in Brasil, and in all or most parts of America, near the sea-shore, and is known by the name of *convolvulus Syriacus*, because it grows in Syria. The root of this plant is long and thick, supplied with nourishment by many small fibrous roots, full of milky juice; from the roots spring large green stalks, which creep along the ground, or climb upon any thing that is near it. Its leaves are green, in the form of a heart; after which come white or purplish flowers, in shape of a bell. The fruit is almost round, and membranaceous, containing black cornered seeds, almost like those of the Spanish arbour-vine. The whole plant is full of milky juice, and smells very strong; which juice is boiled to a consistence. This plant grows in great plenty about Aleppo and St. John de Acre, from whence comes the best scammony. Chuse that which is light, grey, tender, and brittle, being resinous, of a bitter taste and a faint unpleasant smell; reject that which is heavy, hard, and blackish. The next plant that scammony may be got from is,

2. *Convolvulus major polyanthus longissime latissimeque repens floribus albis minoribus odoratis.* Some will have this to be a *mechoacan.*

The *soldanellas* also afford scammony, which purges strongly dropsical humours.

SCORDIUM, *or* WATER-GERMANDER.

The American water-germander, or *scordium,* differs but little from the English *scordium.* It is of a healing and drying quality, and is accounted a good diuretic, alexipharmic, stomachic, pectoral, and vulnerary.

SCOTCH GRASS.

This grass is so called in Jamaica, being brought
hither

hither from a place called Scotland, in Barbadoes. It is a fort of panic grafs, or of the millet kind. This is the only grafs to feed our cattle with; it grows in wet fwampy places, and therefore is green all the year round; fifty acres of it will make more money than any thing we can plant, and is a good eftate.

SEA-FEATHER, *or* SEA-FAN.

I have picked many of them by the fea-fhores and keys; fome blackifh, and fome of a purplifh colour.

SELF-HEAL, *or* ALHEAL.

Thefe herbs are called in Latin *prunella*, or alheal or felf-heal; and the Germans call them *brunella*, or *brunellen*, becaufe they cure that difeafe which they call *die bruen*, common to foldiers in camps and garri-fons, which is an inflammation of the mouth, tongue, and throat, with blacknefs, accompanied with a ftrong burning fever and diftraction or delirium: The juice of thefe plants is a certain fpecific for that diftemper, and all fore mouths and throats, mixed with a little honey of rofes and white-wine vinegar. The decoc-tion of the herb, in wine or water, makes an excellent traumatic drink, to forward the healing of all wounds and ftubborn ulcers. It is faid to take away the pain and fwelling of the tefticles, which negroes are apt to have. Above twenty years paft, one captain Picker-ing, a gentleman I knew very well, had a ftick with fire at the end of it darted at him, which happened to come juft under the brow of his eye, and feemed to turn his eye out, and all defpaired of his life. No furgeon being at hand, they fent for an old negro man, well fkilled in plants; as foon as he came, he ran and took of this herb that hath the bluifh or purple flower, and wafhed it, reduced the eye as well

as

as he could to its place, and then laid on the bruifed
herb, bound it up, and the captain was carried home.
The next day he fent for a furgeon; and, when they
came to open it, found it healed up to admiration;
upon which they fent for the negro, and defired him
to finifh his cure; which he did in two or three days,
only applying the fame thing; and then the captain
rewarded the negro very well, and defired him to fhew
him the herb. This I had from feveral worthy gentle-
men who were there prefent, and affirmed it to be
matter of fact and truth, who fince, they told me, ufe
it to all green wounds with great fuccefs, and call it
Pickering's herb to this day.

SEMPER VIVE.

This is the common aloetic plant which aloes is
made from, and is fo well known in America, where
it grows in great plenty, that there needs no particular
defcription of it.   It is common for planters to give
their children of its thick flimy juice, for worms.
Aloes, which is only the condenfed or infpiffated juice
of this plant, purges and fortifies the ftomach, and is
good againft crude humours, opens obftructions, and
cures furfeits from over eating and drinking; and, if
diffolved in water, and infpiffated again, it fortifies
more and purges lefs.   It preferves dead bodies, heals
and cleanfes old fores.   The Indians have a medicine,
made of myrrh and aloes, called *moceber*, which I have
ufed with wonderful fuccefs in cleanfing old ulcers, and
it will alfo incarnate and heal them if the very bones
were bare, whereas other greafy medicines would foul
the bone; it alfo deftroys maggots or worms in fores,
which are very apt to breed in thefe hot climates. The
juice, drank with milk, heals ulcers in the kidnies or
bladder, and kills worms in man or beaft.   You muft
forbear

forbear giving aloetic medicines to thofe troubled with the bleeding piles, or overflowing of the menfes, to thofe that fpit or vomit blood, or to women with-child. Aloes confifts of two parts, refin and faline ; the one diffolves in common water, the other will not but in fpirit of wine.

SENSIBLE PLANT.

This plant is fo called becaufe, if you touch it never fo lightly, it fhrinks as if fenfible, and folds its leaves clofe together to the middle rib or ftalk, not falling flat down to the ground as the humble plant doth. It hath feveral fmall ftalks and branches from one root, which are hard and woody, with divers joints, at which are little fhort prickled and winged leaves, oppofite to each other, fet very clofe together, and very narrow, fmall, fmooth, and of a frefh green colour. It hath a moffy greenifh white flower.

SEPTFOIL, *or* TORMENTIL.

There is a fort of purple feptfoil, growing about a foot high, on the banks of the river Plate. The whole plant is reftringent.

SHADDOCK.

I have feen them much larger than a man's head. The outfide fkin is of a lemon colour, but very fmooth, and of a fine fcent, exceeding lemon or orange ; its rind is thick, and full of a volatile effential oil ; next the infide fkin is a white fubftance, as in citrons, and then a juicy pulp appears. Thofe of the beft fort are of a deep-red or purple colour ; but thofe that are white are very four, and not good. They fay, if you plant the feed, there is but one in a whole fhaddock that will bring forth good and pleafant fruit : I have

feen

feen many of them planted and come to bear, but never faw a good one produced from the feed. The
beft way is to take a ftem or a twig, and ingraft or ino-
culate it on a good China orange ftock, &c. The fruit
is cooling and refrefhing, abating drought and heat in
fevers.

SILK-GRASS.

This plant is of the aloetic kind. The leaf is not
fo thick and juicy as femper vive, but much longer;
fome are five or fix feet long, but narrow, yet not fo
narrow as the pine or penguin leaf, nor are they fo
broad or thick as the currato. It is full of fmall
prickles on each fide or edge of the leaf, and is tapering from the ground to the top, ending with a fmall
prickle, which makes it of the fhape of a lance.

The chief ufe of this plant is to make filk; which,
as the Indians and negroes make it, is quite coarfe, but
very white, hard, and ftrong; of this they make hammocks and ropes, as alfo fifhing-nets, which will en-
dure the water longer than thread. The way that the
negroes drefs it here, is only to lay the blade, or leaf,
upon a flat piece of wood, and then, holding it faft at
one end, fcrape off, with a blunt lath or piece of wood,
the outward green fubftance, the inward white filk appearing, in ftraight lines or threads, from one end of the
leaf to the other. After they have fcraped both fides,
they throw it into clear water, wafh all the remaining
green from it, dry it in the fun, and then twift it up
into ropes, &c. Undoubtedly, this might be wonder-
fully improved: Nature having fhewn the way, and
brought it to fuch perfection ready to their hands, it
might, with induftry and the art of man, be perfected
much more, to a confiderable profit in making fine
ftuffs of it, and merchandizing in it.

SOAP-

SOAP-BERRIES.

They are fo called becaufe the ciftus or fkins that inclofe thefe berries lather in water, and fcour like foap. When the hollow ciftus or membrane is taken away, there appears a round, fmooth, black berry, of which formerly they made buttons in England. This tree very much refembles the common Englifh afhen-tree in bignefs, colour of bark, and fhape of the leaf; but much differing in the fruit, which is a black round berry, of the bignefs of a marble, contained in a fkin looking and feeling like a dried bladder, very tough, and which doth not ftick clofe to the berry, but feems to have a fpace or hollownefs all round, which is fo tough that you can hardly with your fingers feparate one from the other. Thefe fkins, foaked in water, and rubbed with your hands, will lather and wafh, or fcour, as well as any foap, and have no fmell. The wood is no lafting timber. I have been told, that the afhes of this tree will fpoil a great quantity of other afhes for fcouring or making pot-afh; which feems ftrange, there being fuch a foapy or fcouring quality in the fruit of it.

SORREL.

1. The vine forrel. This with its clavicles lays hold of any thing that it is near, climbing over palifadoes, fo thick that it cannot be feen through, and upon walls, covering them fo that the wall cannot be feen, and keeps green all the year round for many years without decaying. The leaf is thick and juicy, as *orpinant*, or houfe-leek, but much lacerated and divided, fo that one leaf looks like three or four, a little ferrated on the fides, and hath a very four or fharp tafte like forrel, which fome make ufe of for fauce as common forrel,

but

but it is flimy, and leaves a little heat upon the palate. It bears a round berry, like the brionies, firft green, and then very black; when ripe, it hath fometimes a great matted bunch amongft it like dodder, as thick and as big as a man's head; and when it feems to be withered or dried, which this dodder fubftance is, at one time of the year, if you handle or fqueeze it, there will come out a light black fubftance like lamp black, which will ftick fo clofe to the fkin as not eafily to be wafhed off. I believe this might be of ufe for ftaining, colouring, or dying, if rightly underftood.

2. French forrel; of which they make excellent jellies and tarts, not of the leaves of the plant, but of the leaves of the capfula which contain the feed-veffel, and are red, thick, and juicy. Alfo, a fyrup is made of them, far exceeding any fyrup of Englifh forrels: The beft way of making it is to take the red fucculent leaves, and add three times their weight of double-refined fugar; put them together, without water, into a glafs veffel, and then, in *balneo mariæ*, digeft them in a moderate heat, until all the leaves are diffolved, which they will foon be, being foft and full of juice, and make a fine thick fyrup, of a moft beautiful red colour, which will keep much longer than that made with water, and is excellent in fevers, mixed with borage or purflane water. There is alfo a pleafant cool drink made of it with water. The root, given to two drachms, purges very gently the ftomach and bowels.

There are alfo of this fhrub whofe leaves are of a yellowifh-green, as thefe are red, and of the fame ufe and virtues. The bark of this fhrub is very ftrong and tough, like Englifh hemp, and, I believe, would ferve for the fame purpofes.

SOUR-

Sour-Sop.

This is a very common tree in Jamaica, bearing fruit, in shape and bignefs of a bullock's heart, which is very juicy and pleafant to eat. There is a wild fort, called water-apple.

Spanish Arbour-Vine,

Or Spanifh woodbind, which is of the convolvulus tribe. The vines of this plant are fo large and fpreading, that they may be carried over an arbour of an hundred yards long, and that all from one root, which is as large as Englifh briony. It is milky, as is the whole plant, and purgeth very ftrongly all watery humours. I queftion not but a fcammony may be made from it, &c.

Spider-Wort.

There is in America a plant, that grows very plentifully in watery places, like to the Englifh *phalangium,* or fpider-wort. Thefe fpider-worts are all of the fame virtues, and receive their name from having a peculiar quality to expel the bite or venom of fpiders, which, it is faid, they cure infallibly. Some of them grow like water-plantain; fome have a leaf like gentian; fome are branching and fpreading, others not; fome have deep-purple or bluifh flowers, fome have white flowers; another a reddifh or carnation colour; but moft of them foon fade away and fpring again, and therefore have the name of *ephemeras.*

Spikenard.

In America grows, in great plenty, a moft excellent fpikenard. Its leaf is in fhape of the balm, but much bigger, and more like the wild horfe-mint, with

M a large

a large fquare rough ftalk, and globulous head full of fmall blue flowers. It hath a very ftrong fcent, like fpikenard; and if you fqueeze the tops in your hand, a clammy or oily fubftance will ftick to it, and give it a ftrong fcent like the beft oil of fpike. It is an annual plant, and in its greateft perfection about Chrift-mas; in a little time after, none of it is to be feen. It is one of the greateft provokers of urine and ftone-breakers that ever I experienced: I was once fent for to a perfon that lay in a ftrange condition, like hyfteric fits, who, upon nice enquiry, I found was much troubled with the ftone and gravel; and, near upon the time of voiding them, ufed to be fo until fhe voided a ftone or gravel, and then came out of thefe fits; upon which, I ordered a ftrong beverage or fherbet, with lemons, fugar, and a little fpirit of vitriol, and then added an oily fpirit made from this plant, and gave it to her to drink of plentifully like punch, telling them, that if it fuddled her it was no matter, it would do her no harm, for fhe had no fever. She followed my directions, drank plentifully of it, and fell into a found fleep; and, as foon as fhe awaked, made a great quantity of urine, with fmall ftones and gravel; in a few days, there were brought away as many fmall ftones as could be held in the hollow part of one's hand; and fhe was free from thofe fits, nor ever complained of any gravel or ftone, as long as fhe lived after, which was many years. I have often relieved perfons that have had a total ftoppage of urine, and have been in fuch agonies and pain that great fweats and fainting fits have attended them, and death expected every minute, by their only drinking of the aforefaid compofition, which made them evacuate with great violence and in great quantities, bringing away gravel or flime along with their urine, which would fmell very ftrong of the oily fpirit. It alfo expels poifon, and drives out all malignancies.

lignancies. Planters give it decofted to the negroes, to drive out the fmall-pox, and to comfort the heart, as they call it. The dried herb, given in powder, expels wind, cures the cholic, and opens obftructions. The whole plant makes an excellent bath, to take away aches or pains; and heals ulcers.

We have another fort, that is very odoriferous, that grows with a long fpiked head; this I have feen grow to fix or feven feet high; but it is not fo oily as the other fort.

Spirit-Leaf.

This plant is well known in Jamaica by this name. It hath feveral brown and ftraight roots, of an inch and an half or two inches long; from thefe roots arifes a four-fquare ftalk, about nine or twelve inches high, jointed, where come out the leaves, of a dark-bluifh colour; at the top comes out the flower, monopetalous and bell-fafhioned, of a delicate blue colour; after which fucceeds a four-fquare feed-veffel, about an inch long, containing a great many fmall brown flat feeds; which feed-veffel, touched with the leaft moifture, fprings open with a little fnap or noife: And therefore I have advifed a perfon to put one of the feeds in his mouth, and immediately it would fly open, with a leap up to the roof of his mouth, which would furprife thofe who were not acquainted with it. By this fpringing motion, it fcatters its feeds as if fown by art, and often infefts or over-runs great quantities of ground, not to be got out without much pains and difficulty. The whole plant much refembles the *gentianella alpina verna major et minor* of Parkinfon. It is an admirable vulnerary herb; the planters make an excellent balfam of it, to cleanfe and heal all ulcers. It is alfo called *felwort*.

M 2   Spleen-

Spleen-Wort.

Thefe are of the fern kind. They are accounted fpecifics for all diftempers of the fpleen, wherefore they have the name of fpleen-wort; they open obftructions, and therefore good againft the yellow jaundice; they take away hiccoughs and ftrangury, expel gravel, and help a violent gonorrhœa.

Spunges.

We often meet with thefe on the fea-fhores of America.

Spurges.

There are many forts of fpurges growing in America, even from a tree to one of the fmalleft creeping vegetables.

1. Called *tithymalus arbor Americana mali medici foliis amplioribus tenuiffime crenatis fucco maxime venenato.* This is very venomous.

2. *Ricini fructu glabro arbor julifera lactefcens folio myrtino.*

3. *Thymelæa maritima ericæ foliis furculis tumidis et tomentofis,* which is a fort of fpurge-olive.

4. *Thymelæa humilior foliis acutis atrovirentibus.* Thefe are violent workers upwards and downwards, and therefore ought to be difcreetly given.

5. *Cajatia,* alias *caacica.* The Brafilians fet a very great value upon this plant. Pifo faith, it is one of the beft antidotes in the world to expel all forts of poifon; even, faith he, when it hath reached the very heart, which it corroborates and fets a-beating, when juft leaving off its office of pulfation, and caufes the blood to circulate again, and that by only giving a pugil of the dried herb in a proper vehicle, or by giving

the

the juice of the green herb; alfo, the herb decocted, or infufed in wine, doth the fame. The green herb, bruifed and applied as a poultice to the part bit or ftung by any ferpent or venomous creature, it immediately takes away the pain, and draws out the venom, preventing it fpreading all over the body of fluids: From experience, faith Pifo, one drop of the juice of this plant, dropped upon a ferpent, immediately kills it; and for that reafon, there is no prudent perfon, that goes in the woods of Brafil, will go without fome of this herb. A bath made of the whole plant, with cotton-tree bark, takes away carbuncles and phlegmons. It is alfo experienced to be excellent in all venereal cafes, as alfo a fpecific remedy in the belly-ache, as you may fee by Dr. Trapham's account of it, in his State of Health of Jamaica; where he fays, " As for a fpecific for the dry belly-ache, take an Indian one (for the Indians have many fuch), which my worthy friend and fagacious, Dr. Lawford, of the ifland of Barbadoes, communicated to his excellency Lord Vaughan, by whofe favour, for the benefit of the afflicted," faith Dr. Trapham, " it was communicated to me: The faid Dr. Lawford affirmed, that he had had above one hundred trials of this plant, of which, faid he, I give a drachm of it, powdered, in any convenient liquor, and repeat it, once in three or four hours, till the ufual fymptoms of the difeafe abate; fometimes, I give it made into a fyrup, of which I give one ounce to three; alfo, in decoctions and clyfters. It is alfo, faid the fame doctor, an antidote againft poifon, and a great diaphoretic, expelling all malignancies in fevers." Trapham faith, the Englifh in Barbadoes called it fnakeweed; " and," faith he, " after the fymptoms of the belly-ache are removed by this fpecific plant, I would

have

have them apply a plaifter of hog-gum to the weak limbs, ufing warm frictions, and renewing the plaifter every twenty-four hours, which reftores the ufe of the limbs," &c. -

6. Sir Hans Sloane calls *tithymalus erectus acris pa-rietariæ foliis glabris floribus ad caulis nodos conglome-ratis.* This is not of the fweet nature as the former, and yet more fafe to be taken inwardly than the reft of the common fpurges, but muft be ufed with difcretion.

7. The other is a fort of thyme, the fmalleft fpurge of them all, and the moft common, for it grows every where, even in the ftreets, between paved ftones and bricks. I have known feveral perfons ufe it, with good fuccefs, to take off the fpots or films on the eyes, that have come after the fmall-pox, and that by only drop-ping the milky juice into them; but I fhould think it more fafe to mix it with a little honey, for it eats off all forts of warts. The people in Jamaica call it eye-bright, for its great cures to the eyes.

Spurges are generally of one and the fame kind, only fome more violent in their operation than others, except the fweet fpurge called *caiaca*, mentioned be-fore, which hath a quite different nature; for, as all other fpurges work upwards and downwards, this doth neither, but operates by fweat and urine. The rea-fon of the others working fo ftrong, is from their abound-ing with an effential fixed acrid falt and oil, and there-fore dangerous to be adminiftered without correcting; but, when corrected, they may be given with fafety in dropfies, lethargies, phrenfies, &c. You may make an extract of them, which fome ufe as a general purger. Raius faith, that fpurge-laurel, powdered and infufed in wine-vinegars, cures cancers.

STAR-APPLE.

The fruit of this tree is as round as a ball, as big as the largeft of our Englifh apples, and, when cut acrofs, refembles a ftar, the feed partition making it fo. It hath a thin fkin, containing a foft pulpy fweet fub-ftance, but clammy; the ftones, or feeds, are almoft in fhape of a prune ftone, and nearly as hard, partly fmooth and partly rough. They are fine large fpread-ing trees, whofe leaves are in fhape and bignefs of the cafhew, but of a fine deep-green on the upper fide, and of a fine foliomort underneath. They bear but once a-year, which is about Chriftmas, and then their fruit is fold about the ftreets, and much admired by fome.

STAR-STONES.

We meet with feveral of thefe ftones by the fea-fide. They are of the coral kind. Some are called brain-ftones, becaufe upon the furface of them there is a reprefentation of the meanders, windings, and turn-ings, which appear upon the furface of the brain. Some have fhining fpecks in them, like ftars; and others are called rofe-ftones.

STAR-WORT.

There are feveral forts of thefe plants. Plumier defcribes feven forts, and Sir H. Sloane two. Star-worts are recommended for the cure of venereal tu-mours, as alfo to abate inflammations of quinfies in the throat, and cleanfe and heal ulcers there or elfe-where; to ftop defluxions of all humours, and good in inward bruifes. Craterus commends a decoction or fyrup of the flowers, to cure the falling ficknefs. The feeds are aromatic; and Pifo fays, the whole herb makes an excellent bath for pains and inflammations.

M 4 STOCK-

Stockvishhout.

The Dutch give this wood, that grows about the Lagoon of Nicaragua, the name of *stockvishhout;* but with us it is commonly called Nicaragua wood. It is but small to what logwood is, seems to be very tough, and is about the bigness of dried stockfish; which may be the reason the Dutch call it *stockvishhout.* It dyes a very fine red.

Stoechas.

We have a wild sort or two of *stoechas:* One sort is called by some *caffidony,* or French lavender; another is a sort of cudweed. These plants are very astringent, and therefore proper for fluxes of the body, and all defluxions of rheums. A syrup made of the tops of it, when in flower, is good for coughs and catarrhs.

Strawberries.

English strawberries will grow in America as well as in England, if care be taken of them: They are apt to spread themselves in strings and runners, covering great quantities of ground; and then they will blossom but not bear well. In Chili, they plant whole fields with a sort of strawberry, much different from ours (the leaves are rounder, thicker, and more downy), which they call *frutilla.* The fruit is generally as big as a walnut, and sometimes as an hen's egg, of a whitish-red, but not so delicious as our wood-strawberries, and more of the taste of the English little hoboy-strawberries.

Sun-Flower.

These grow as well and as large, or rather larger than in England; they are the very same sort, and have all the same virtues.

SUPPLE-JACK

Is a withe fo called, which is full of round knobs at every five or fix inches diftance, and, for the toughnefs and fupplenefs, called fupple-jack. They are of all fizes; but if you do not oil them now and then, they will grow very brittle, &c.

SWEET-SOP.

A leaf, laid on pillows or beds, will draw all the chinks or bugs to them, fo as you may be rid of them.

SWEET-WOOD.

Befides the lignum aloes and rhodium, we have another fweet-fcented wood, commonly called timber fweet-wood-tree, which is of the laurel-leaf kind. At one time of the year, the pigeons feed upon the berries of thefe trees, and then their inward parts, and fometimes their flefh, is very bitter.

SYCAMORE.

I have often feen, as I have rode along, a fmall plant among the bufhes, growing about fix or feven feet high, which feemed not to be able to fupport itfelf; but yet did not climb about any thing: It had a very fmall ftalk, and but few leaves, as large as a laurel, but thin and fofter. At the top were branches of yellowifh flowers; afterwards came winged feed-veffels, exactly like the fycamore.

TACAMAHAC.

This gum flows from the bodies of large thick trees, like the Englifh poplars, growing plentifully in New Spain and Madagafcar, where they are called *hazame*. The leaves are fmall and dented, the fruit red, of the

fize

size of a large nut, with a peach-like stone. *See the figure of it in Pijo.* It is said to ease all manner of pains in the head, nerves, joints, or womb, and to be very good in vapours. It is so famous among the Indians in America, that they use it in all pains whatsoever. It much resembles *galbanum*.

TAMARINDS.

The tamarind-tree is well known in Jamaica. The pulp of this fruit is purging and very cooling, quenching thirst, and abating the heat of inflammatory fevers; the only fault is, it is apt to gripe some persons violently. It opens obstructions, and is good against pimples or breakings-out, which proceed from the heat of blood and salt humours; with borage-water, it is excellent for heat of urine proceeding from a venereal cause; and is a very good purge, mixed with rhubarb and caffia, for the same distemper.

Here are also two or three wild tamarind-trees, but the fruit is of no use; their bodies are excellent hard timber: The one is called red, and the other white, tamarind; another sort hath leaves like *tamarisk* or *savin*, but its fruit unlike, which is an excellent restringent to stop fluxes of blood, and cleanse and heal old ulcers.

TAYO.

This is a large sort of *eddo*. The roots of these, although never so well boiled, will heat the throat (which is called scratching the throat), and therefore are generally given to hogs to eat.

TEA.

And first, that which is called Lima tea. Feuille saith, the virtue of this herb (which is the same with the

the China tea) was not known in Peru till 1709: Then we in Jamaica were beforehand with them, for it hath been known with us above thirty years; and about twenty years paſt, a French captain of a ſhip affirmed to me, as we were walking together about our town called St. Jago de la Vega, obſerving this plant grow in ſuch plenty, that it was the very ſame plant as that of China tea, and that he lived there many years, and had ſeen large fields of it, and the way of cultivating it; and all the difference was, theirs was larger, owing to their cultivation. This plant, Sir H. Sloane takes notice of in his Hiſtory of Jamaica, and makes it a ſort of hedge-hyſſop. Paul Hermanus calls it *capraria Curaſſavica*, from the Dutch in Curaçoa, who call it *cabrita*, from the goats feeding upon it; but I never ſaw the goats in Jamaica eat of it. It is called the leſſer tea. Now, to clear the doubt whether this be the ſame tea that grows in China and Japan, I will firſt deſcribe the plant which is called Weſt-Indian tea, and then the Eaſt-India tea, by which we may ſee the difference. And firſt, obſerve that this plant hath ſeveral ſmall long brown roots, about two inches long, which ſend up a ſtem three or four feet high (and would be much higher, if cultivated), woody, covered with a ſmooth clay-coloured bark, and having ſeveral branches, which are very thick ſet with leaves, without any or-der; each leaf is from one inch to two inches long, and about half an inch broad about the middle (where it is broadeſt), and then goes off tapering to a ſharp point, having no foot-ſtalks, of a deep-green colour, ſmooth and thin, being a little ſerrated on its edges, and they would be much larger if planted in good ground, and taken care of as they do in the Eaſt-In-dies. Between the leaves and ſtalk of the branches ſome the flowers ſtanding on a ſhort foot-ſtalk, which

are

are fmall and white, feeming to have five leaves, but
are only deeply divided into five parts, ftanding in
a green calyx; then comes the feed-veffel, which is
oblong, cylindrical, four-cornered, but very fmall, and
when dry is of a light-brown colour, in which are contained a great many fmall brown feeds.

Now, to come at the true knowledge of the China
tea is no fmall difficulty. Bontius hath pretended to
give us a figure of the plant, which feems to differ
very much from the defcription of others, of this plant;
and for the better clearing and paffing a judgment upon
this plant, I fhall fet forth the feveral different accounts
of it, and fhall begin firft with Bontius.

The annotators upon Bontius fay, it is no wonder
if, about this noble Afiatic plant, there fhould be fuch
different accounts, the natives having fo referved it as a
fecret to themfelves, fuffering no ftranger to fee it growing; and if any afked them whence it came, and where
it grew, they would always prevaricate, and never an-
fwer directly: Sometimes they would call it an herb,
another time a fhrub, fo that nothing of certainty could
be concluded from what they faid. As to the figure
and manner of its growing, Bontius himfelf complains
he was never able to obtain; but at length, by the favour of Lord Caron, a worthy prefect of Japan, he
obtained a draft of the plant, which he hath given us,
which fhews the bignefs of the leaves, flower, and feed,
which indeed differs little or nothing from that which
grows with us in America, only the leaves are broader,
and the whole plant larger, which may be owing to
their cultivation. The figure of the plant having not
hitherto been given, until Bontius gave it us, it is no
wonder that many have erred about this plant, and
feem to make a difference between tea and *tfia*, when
they are both one and the fame plant, the Chinefe
calling

calling it tea, and the Japanese *tſia*. This ſhrub, ſaith
Bontius, is of the height and bigneſs of our European
currant-buſhes; the ſtalks and branches, from the foot
to the top, are adorned with tender pointed leaves and
flowers, which are very numerous, whoſe leaves, although of the ſame form, yet are ſo different in bigneſs that, upon one and the ſame ſhrub, are accounted
five different ſorts of tea; *viz.* the firſt and loweſt
leaves, neareſt the bottom of the ſhrub, are the broadeſt;
to theſe ſucceed a ſecond ſort, much ſmaller, and ſo on
to the top; and the ſmalleſt leaf is accounted the beſt.
The difference of leaves is no more than what is common
to many plants, and is the ſame with our American tea.
The flowers are in bigneſs, colour, and ſhape, like
our eglantine or ſweet brier, but not in ſmell. The
flower falling off, there remains a fruit like a navel,
containing a round black ſeed (herein it differs vaſtly
from the American tea). The root is fibrous, diſ-
perſed in very ſmall fibres into the ground, to draw its
nouriſhment. The leaves, when green, are ſomewhat
aromatic, beſides a little hottiſh and bitteriſh (herein it
differs much from ours). Some have teſtified, and it
ſeems moſt probable to be true, that this plant at firſt
grew wild in China, and lay long neglected, but by
its culture, high virtues, manner of preparing, and
daily uſe, is but modern as it now appears; and ſo
were tobacco, ſugar-canes, and indigo, which in former
times were wild, and not minded till the qualities of
them were diſcovered by the native Indians to people
of other nations, and then they were ſoon improved
by cultivation, with pleaſant and profitable tranſport-
ation through all the known world; and who knows
what perfection our wild tea might be brought to, if
the ſame pains and labour were taken with it as in the
Eaſt-Indies? But I ſhall now proceed to other ac-

L.

counts of the Eaſt-India tea; and the next will be Breynius and Ray's account of it;

Who ſay, that tea, or *tſia*, is a ſhrub, with many thick branches with dark-green leaves, jagged on the edges like a ſaw, being in ſubſtance and form more like the ſpike-willow of Theophraſtus than the ſweet willow, and of a drying taſte, with ſome bitterneſs. The flowers are white and five-leaved, and about the bigneſs of the female ciſtus, having many thrums in the middle; to theſe ſucceed the fruit, which is green when freſh, but when dry is covered with a dark-brown ſkin, and the ſhape as variable as the Eaſtern coccus, ſome roundiſh, and of that bigneſs, or of ſpurge-laurel, others twice as long, containing one ſingle ſeed, others two, and ſome three ſeeds, the huſk being parted into ſo many cells or partitions (not unlike the ſmall *ricinus*), which when ripe opens and turns out its ſeed, which are roundiſh, of a tender and light ſubſtance, and of a browniſh colour.

The next is Monſieur Pommet's account of this tea. His figure of the plant differs much from that of Bontius, both in leaf, which he makes much broader, and the fruit or ſeed-veſſel, which he makes a perfect tricoccos. Indeed, he ſaith, it hath a ſlender green thin leaf (but his figure is not ſo), pointed at the end, and a little ſerrated or jagged; after the leaves come ſeveral cods, of the bigneſs of the end of one's finger, in ſhape like the *areca*, in each of which are two or three berries, of a mouſe-coloured grey without, and within having a white kernel, very ſubject to be worm-eaten; but takes no notice of the flower.

Monſieur Lemery ſaith, that tea, or *tſia*, hath a ſmall fibrous root, ſending forth branches beſet with ſmall, oblong, ſharp-pointed, thin, green leaves, a little jagged or ſerrated on the edges; the flower is white and

and pentapetalous, formed like a rofe, with fome ſtamina or thrums, which, when gone, is fucceeded by a little cod, like a fmall hazel-nut, of a chefnut colour, containing two or three kernels of an almond ſhape, but ſmaller, and of an ill or difagreeable taſte.

So that, by all theſe different accounts, this Eaſt-India tea agrees with the Weſt only in the leaf and colour, and divifion of their flowers; but then the Eaſt-India hath a large flower, whereas the Weſt-India hath a very fmall one; then, as to the feed-veſſel, they altogether differ fo much, that it can never be one and the fame plant, although it may have the fame virtues, which are very great, if you believe them that write of it: But if the profit in merchandizing of it were not more than its virtues, it would foon be brought into difefteem. One great reafon of tea becoming fuch a commodity throughout all Europe is, becaufe the Dutch change it for fage, which the Japanefe and Chinefe are great lovers of, which certainly is more medicinal and of more value than their tea, and what they themfelves are not infenfible of, which makes them prefer our European fage much before their own tea, and wonder at the fame time we fet no greater value or efteem for it. I knew a gentleman in Jamaica who drank no other tea than what grows with us, and although he could not curl it up fo artificially, yet he did it pretty well; and all that he treated with it praifed it to be the beft green tea they ever drank in their lives; and I am of that opinion, for it hath as many virtues. In the fame manner, about forty years paft, I knew a gentleman at Norwich, who ufed to treat the ladies with tea, and they would fay, " Lord, Sir Thomas, you have the fineft tea in the world! it hath fuch a fine flavour! pray where do you get it?" " Oh, ladies, that is a fecret!" Afterwards, he ferioufly told me, and avouched it for a truth, his tea was only new hay.

THORNY

THORNY APPLES.

There are three forts of this plant. One hath a very white flower. Of this fort I faw growing in a garden in Colchefter, above forty years ago: The furgeon who had it made both falves and ointments of it, the ufe of which gained him much credit; and there is an account in Gerrard, of a gentlewoman in Colchefter, who was fo burnt with lightning as to be thought paft all relief, but was cured by an ointment made of the leaves of this plant. I have known it experimentally cure contracted tendons or nerves, by chafing or rubbing in the ointment hot into the part affected. It hath a thin green ftinking leaf, fmelling almoft like opium, and much indented; it branches and fpreads like a little tree; the ftalks are of a pale-green; it hath a long tubical white flower, after which comes its fruit, which is oblong, and in fhape and bignefs of a walnut with its green fhell, fet full of foft prickles while green, but when dry are able to penetrate into the flefh; thefe contain a vaft quantity of fmall black feeds, like the *papaver fpinofum*, and of a ftupifying quality. I know a gentleman at this prefent time, that, whenever he hath a fit of the gout, applies thefe leaves to the part, and it gives eafe in about three hours. The leaves, applied to the head, eafe pain and caufe reft.

There is another fort, commonly called trumpet-flower (becaufe it is fo long and large, in the fhape of a trumpet or hautboy), of a fine purple colour without-fide, a fine white within, as foft as velvet, and of a delicate fweet fcent; fome of them are double-flowered like a trumpet; all its ftalks are black and fhining; its fruit or feed-veffels, inftead of prickles, are full of little protuberances like warts; its feed is flat, and of a light-brown colour.

The

The third fort hath the fame kind of feed-veffel, but a little more prickly; its flalks are of a fhining black, its flowers of a pale-blue, but not fo long nor fo large as the former.

THOUPA.

This is a fhrub like horfe-tongue. The flower of it is long, of an Aurora colour, refembling that of birth-wort; from its leaves and rind proceeds a yellow milk, with which they cure ulcers; but fome will have it to be a poifon-plant. It grows in Chili, and moft fouth-ern parts of America.

THROAT-WORT.

This plant is fo called from its fpecific quality in curing difeafes of the throat. It hath fome refem-blance to the valerian. There is very little of it grows in America.

TOAD-FLAX.

There are feveral of thefe kinds of plants in America.
1. American toad-flax, with a fmall yellow flower.
2. *Linaria paluftris fœniculæ folio*, of Plumier.
3. *Linaria minor erecta cærulea*, of Sir Hans Sloane. It hath a round fingle ftalk, rifing about two feet high, on which are placed leaves alternatively, about an inch and half long, but narrow, like the leaves of *linaria lutea vulgaris*. The tops of the ftalks are branched into feveral long fpikes, fix inches long, full of blue flowers as the others of this kind, after which follow fo many roundifh turgid feed-veffels, each di-vided into two cells, in which lie flat brown feeds. Thefe have much the fame virtues as common flax-weed; the juice, mixed with hogs lard, is a moft ef-fectual remedy for the pain and fwelling of the piles or hæmorrhoids.

TOBACCO.

TOBACCO.

The juice of green tobacco deſtroys maggots in ſores beyond any thing that can be made uſe of; it makes an excellent healing balſam or ſalve; an oil, drawn in a retort from dried tobacco, ſcales the bones, cleanſes the fouleſt ulcers, and takes away their callous edges, making them fit to heal; the white aſhes cleanſe the teeth, and kill the worms in them.

TOOTH-WORT.

This plant is ſo called from the form and colour of the root, which is very white, and is compoſed, as it were, of a great many teeth. We have a ſort of it growing in America; ſome will have it to be a ſort of lead-wort. This plant hath a viſcous green calyx, in which is a white pentapetalous flower, like the *lychnis ſylveſtris flore albo*, with a rough viſcid capſula, which catches flies. This plant is not a true climber, and yet it cannot ſupport itſelf, it generally growing amongſt ſhrubs. It is counted a cooling, drying, and reſtringent plant, therefore good in ruptures, and a good vulnerary herb for wounds: Some make it to have the properties of wild campions, others of lung-wort.

TRAVELLERS JOY.

This is a great climber. I never could underſtand why it is called travellers joy, or what joy travellers reap from it: The country people in England call it *honeſly;* but we never make any uſe of it but to tie rails with, and it is commonly called pudding-withe, being ſoft and pappy whilſt green, and of a hot biting taſte. The juice and flowers, beaten and boiled, and then rubbed and applied on the ſkin, take off ſpots and freckles; the root, infuſed in ſalt water, and mixed with wine, purges all watery humours.

TREE-

TREE-ROSEMARY.

This I happened to meet with by chance. Pulling down fome old houfes, I fmelt a very ftrong fmell of rofemary, which made me enquire into the reafon of it. They told me, that there was fome rofemary-wood among the timber of the houfes. I then defired they would get me fome of it, which they did; I found it was only the bark that fmelt; which no rofemary exceeded. Some will have it to be a fort of clove-bark tree, which grows in great plenty upon the main continent. I firft found this tree on Bachelor's Plantation, which was afterwards mine; and is now well known to all or moft planters in Jamaica. I carried fome of the bark with me to England in the year 1717, which kept its fcent very well; and I queftion not but it would be found, upon experience, to be very ufeful to diftillers, and of many medicinal ufes.

TRUMPET-TREE.

This is the common name this tree is called by in Jamaica, I fuppofe from its hollownefs. It bears a long, crooked, foft julus, reprefenting or refembling worms, and hath a very large indented leaf. It is of a very quick growth, growing very ftraight and tall, without any branches, and at the top there is a foft pappy fubftance, which fome will eat; cattle will eat the leaves and its fruit, fo will pigeons. The holly on the top of the tree contains a white, fat, and juicy pith, which fome eat; but the negroes, with this, and with the young tender foft leaves, cure their wounds and old ulcers. I was once in the woods, and was caught in a great fhower of rain, having only an old Congo negro with me, who made me a hut; and I, having heard that fome negroes could make fire, as they called

N 2   it,

it, I afked him if he could do it; he faid yes, and
went and got a drv piece of this tree, and fplit it, making a little hole or dent in it with the point of his knife;
he then took a fmall piece of harder wood, and made
the end of it to fit that dent; then he fat down, and
held the flat piece between his feet, and with the
upright piece, which centered in the hollow of the
other, twirled it round very fwift between the two
palms of his hands; it began to fmoke in a very
little time, and fire appeared, which he fo managed
that we had foon a very good fire. The juice of the
tender tops is aftringent, and good againft fluxes, immoderate *catamenia*, and gonorrœas; it is alfo good
againft the immoderate *lochia*, if a poultice of the
leaves be applied to the navel. Its bark is very tough,
and makes as good ropes as thofe of hemp. I knew a
phyfician that cured many dropfical negroes with the
afhes of this tree, which afterwards I made ufe of for
the fame purpofe; and I obferved, that they were the
heavieft afh that I ever faw (which I difcovered by
weighing them with other wood-afhes), and made a
ftronger lixivium than any others, having a greater
quantity of fixed falt in them; they are therefore proper for dropfical perfons.

TURNSOLES.

Thefe plants have their names from their flowers
always turning to the fun, and are called from the
Greek *heliotropiums*. There are feveral kinds of them
in Jamaica.

1. *Heliotropium arboreum maritimum.* This plant
hath fucculent or thick juicy leaves, covered over with
much white down, like the American cudweed or cot-
ton-weed; the tops are branched out into feveral fpikes
of white flowers, contorted like a fcorpion's tail.

2. A

2. A fort of wild fampier, defcribed already.

3. A greater fort, with a white flower.

4. The wild clary.

5. Another fort, with narrower clary leaves.

6. *Heliotropium minus lithofpermi foliis,* a fort with a groundfel leaf. It cools and gently purges by ftool, and is counted a fpecific againft the poifon of the *phalangium* fpider, and againft fcorpions.

VALERIAN.

We have very little or none of the true valerian growing in America, that ever I could find. Sir H. Sloane takes notice of two forts of valerianellas: the firft is called hog-weed, mentioned before. Of the fecond fort, the lower part of the main ftem is as big as one's arm, having a furrowed white bark. It is a climber, taking hold of any palifadoes or trees it comes near, and branches at the top, rifing feven or eight feet high. The branches are many, round, red or green, and brittle, hanging downwards; the leaves come out at the joints, and are exactly like thofe of the greater fort of chick-weed; the tops of the twigs fend out feveral *radii,* or foot-ftalks, as from their common centre, like the *umbelliferæ,* fuftaining each one fmall greenifh-yellow flower, like a fmall cup, being round, undivided, and almoft like *mufcus pyxydatus* in fhape; after which comes a fmall, long, brown feed, almoft like thofe of fome *umbelliferæ,* growing longer from the beginning to the top, and being a little tough. It is a cooling and moiftening herb. It grows in moft hedge-rows and fences every where.

VANILLAS, or BANILLAS.

This is a convulvulus plant, climbing about fhrubs and trees. The fruit or pod is called by the Indians

in

in Mexico *mecafuthil*. I have feen it grow in Jamaica, but I never knew any perfon there that could cure it, or bring it to its fine fcent, as the Spaniards do at Campeche and Bocatoro Guatulco. It is a vine, with a round, jointed, yellowifh-green ftalk, putting forth here and there leaves of the bignefs and fhape of the velvet-leaf; its flowers are of a whitifh-yellow colour, almoft in fhape of a hand; after the flowers follows a flender long pod, five or fix inches long, full of fmall feed; the outfide fkin of the pod is firft green, and when ripe yellowifh, and, as they dry and are cured, grows black and fhrivelled; upon opening it, the feeds within are fo thick, fmall, and fine, that it looks like velvet. Although they grow in great plenty in moft parts of America, yet few know how to cure them, fo as to have their true aromatic fcent, the Spaniards keeping the fecret to themfelves; but the Indians, who taught them, informed me, that they had but two ways of curing them: The firft was, when they are juft ripe (for if you let them ftand too long they will fplit open of themfelves, in order to fcatter their feed, and then you can never cure them) they gather them, and hang them up by one end, in a fhady place, to dry; as they dry, they now and then prefs them gently between their fingers, which makes them flat, and then flicken them over with a little oil, which hinders them from drying too faft and fplitting open; and fo repeat, until they are fit to be rolled up neatly in papers. The other way is, to gather them as before, and fcald them in the following liquor; *viz.* Make a brine with falt and water, ftrong enough to bear an egg; then put in a fourth part of chamber-lye, and a reafonable quantity of quick-lime, which mix, and boil together about half an hour; then take it off, and put your vanillas into the liquor; let them remain there until they are thoroughly fcalded

or

or parboiled; then take them out, and dry them in the fhade, or where no fun can come to them. In the fame manner you may cure China-root; but inftead of drying it in the fhade, you muft dry it in the hot fun, and then no worms will take it; and if a little of the red colour comes out of the root it is never the worfe for fale (fo that you can but keep the worm from it), for the paleft china-root is now become the moft valuable. The Spaniards have a particular way of manuring and cultivating the grounds where they plant their vanillas, or otherwife they would make but little advantage of them, as the Japanefe and Chinefe do their tea; and, after planting them in well-dunged land, they take care to mould them up as they grow, and then put poles for them to run upon, as we do hops in England; then they take them juft in their full ripenefs and fcent, they having a moft particular odoriferous fcent, and yielding a great deal of oil and volatile falt. They are very cordial, cephalic, ftomachic, aperitive, and carminative, opening all obftructions, attenuating vifcous humours, provoking urine and the menftrual difcharge. It is often mixed by the Spaniards with their chocolate, which makes it have a pleafant fcent; and then, to make it of a fine yellow or golden colour, they add anotto, the Spaniards having a very great opinion of its virtues. It is fuppofed, that the fcent in Warham's apopletic balfam, for which he got a patent, was owing to vanillas, &c.

VERVAIN.

We have feveral forts of thefe plants. One fort is exactly like that in England; it keepeth green all the year round. This fort is well known by moft or all the inhabitants of America: The Indian and negro doctors perform great cures with it in dropfies, efpe-

N 4

cially

cially thofe in women, occafioned by obftructions of the menftrual difcharge, and that by only giving the juice of the plant. It is a powerful remedy againft worms, as was evident by a gentlewoman in America, who was in a lingering confumptive condition for fome time, and the occafion of it could not be found out by the phyficians: Her lungs were good, and fo was her appetite, but fhe ftill wafted, and was always complaining; at laft, a fkilful Indian gave her the juice of this plant, mixed with fome fugar, by the ufe of which fhe voided, in a few days, a thick worm, above twelve inches long, hairy, and forked at the tail, after which fhe foon recovered, and was perfectly well. The fame perfon recommended this remedy to another gentlewoman in Peru, who, by taking it in the fame manner, voided many fmall long worms, and, among the reft, one very long and flat, like unto a long white girdle; after which fhe alfo became well. It is almoft certain, that the death of moft children in America is occafioned by worms, entirely owing to their fruit, which is very apt to breed them: This might be often prevented, by taking the juice of this plant, with contrayerva infufed in wine; which would alfo prevent the fever that is occafioned by them. The ancients attributed many virtues to vervain: It is a great cephalic, and vulnerary in the diftempers of the eyes and breaft, in obftructions of the liver and fpleen; it makes an excellent gargarifm for difeafes of the throat, and is good againft piles and falling-down of the anus.

To take away the hardnefs of the fpleen, *bruife vervain with the white of an egg and barley-meal or wheat-flour; make it into a cataplafm, and apply it to the part.*

VELVET-LEAF.

This is a convolvulus plant. It grows in great plenty amongft amongſt ebonies, climbing about them. Its leaves are as ſoft as any velvet, which makes the planters call it velvet-leaf; they are about the bigneſs of an Engliſh crown piece, rounding like the *aſſarabacca, &c.* of a yellowiſh-green colour. It is a moſt excellent antidote againſt poiſon, inwardly taken or outwardly applied; I have ſeen it heal a wound to admiration, by juſt laying one of the leaves upon the wound; it cures ulcers in the lungs. I knew a phyſician perform great cures on conſumptive perſons, who told me that his remedy was only a ſyrup made of the leaves and root of this plant, for which he had a piſtole a bottle.

VINES.

There are ſeveral ſorts of wild vines in America, bearing fruit.

1. Thoſe that climb upon trees, and have a very pleaſant, ſmall, black grape. [*See* Water-Withe.]

2. The wild vine of Virginia.

3. The wild vine of Canada.

Wild vines are of the ſame nature, virtue, and quality, as the manured, which are pleaſant to the ſtomach, and provoke urine; the leaves make a good mouth-water, and an excellent bath or waſh for the piles, *&c.* The aſhes of the branches clear the eyes of films, ſores, and ulcers, and take away the overgrowing ſkins of the nails of the hands and toes.

VIOLETS.

We have ſome plants whoſe flowers reſemble European violets, but come ſhort of their fragrant ſmell; as,

1. The tall Chili violet, without ſcent, but its flowers of a deep-blue; of which they make a tea which is very opening.

2. The creſs violet of Peru. This elegant plant the
Spaniards

Spaniards call *paxaritos*, becaufe its flower is compofed of two particular large yellow leaves at bottom of the flower, extended like the wings of a bird. It grows about Lima.

3. Sir H. Sloane's *violæ folio baccifera repens flore albo pentapetaloide fructu rubro tricocco.* This herb has a fmall, round, creeping ftem, putting forth at its joints many fmall fibrous roots, and having fmall branches at about an inch diftance from one another, each of which is about an inch and a half long, having roundifh leaves ftanding oppofite to one another, on an inch-long reddifh foot-ftalk, in every thing refembling thofe of violets, only fmaller and rounder. The flowers come out at the tops of the branches; they are white, and divided in their margins into five fections; then come feveral round fmooth berries, as big as an Englifh pea, containing, in an orange-coloured pulp, two long brown feeds. It loves to grow in fhady moift places, by the fides of woods. The berries, or whole plant, boiled in whey, cure fluxes; and, boiled in oil, cure blood-fhot eyes.

4. The corn violet, dame's violet, and Venus's looking-glafs. It puts out its flowers a little before Chriftmas with us in America; they are of a fine blue colour, with five fections, making a fine fhow, like blue pinks. It grows almoft every where in America: The whole plant is hot and dry in the third degree, and much of the nature of rocket; the diftilled water of the flowers, inwardly taken, caufes fweat, and, outwardly, is a good beauty-wafh.

VIRAVIDA,

Is the name they give a fort of femper vive in South America; the infufion whereof was ufed with great fuccefs by a French furgeon, for curing a tertian ague.

VIRGINIA

Virginia Snake-Root.

This is called *polyrhifos Virginiana*, or the rattle-fnake weed of Virginia.

Wake Robin, *or* Arums,

Of which there is great variety.

1. The *tayas*, mentioned before.
2. The lesser *tayas*.
3. The *eddos*. Thefe three are eaten as bread-kind, as fhewn before.
4. The dumb-cane, mentioned before.

The roots of every fpecies of thefe plants, but efpe-cially of the fpotted ones, have an extraoidinary acri-mony, fo that if you tafte any of them, they will bite your tongue the whole day : But how biting foever they be, if their roots are thoroughly dried, and kept for fome time, they lofe all their acrimony, become infipid in tafte, and may be taken very fafely. The dried root, pulverized and mixed with honey, power-fully expectorates thick and tough matter, and is therefore excellent in afthmas. The roots of arum are the bafis in the ftomachic powder of Quercetanus. A drachm of the root in powder, given in a proper ve-hicle, is an excellent remedy againft the plague or pef-tilential fevers, and againft poifon ; taken in white or Rhenifh wine, provokes urine, brings down the monthly purgations, purges effectually of the *lochia*, and brings away the after-birth; taken with fheeps milk, helps inward ulcers; the frefh roots and leaves diftilled, with a little milk, make a fine beauty-wafh, and is an excellent water for all forts of fpotted and malig-nant fevers; the powder of the roots, mixed with flour of brimftone, is a fovereign remedy for a con-fumption; the root bruifed, or the leaves, applied as

a poula poultice, ripens any boil or plague-fore; the juice of the leaves cures a polypus in the nose, and all foul ulcers.

Besides the arums, there are several American dragons or dracunculufes: 1. The American dragon, with snipped or jagged leaves, which, upon each knot of the stalk, sends forth two roots from each side, which stick close, if not insinuate or penetrate, into the bark of the tree; the foot-stalks of the leaves are longer and thicker than those of the *colocasia hederacea ster lis latifolia*, and the leaves near to the same size, thickness, and colour, deeply divided round the edges, like the *palma Christi*; from the middle nerve or rib of the leaf there is a pretty thick nerve, that reaches to the extremity of each fegment. Its leaves bruifed, and mixed with hogs lard, make an excellent unguent for old ulcers in legs; which, Dampier faith, one of their ship's crew learnt from an Indian. They are of the nature and quality of arums, but in a leffer degree of heat and pungency.

WALL-FLOWERS.

Sir H. Sloane, in his Natural History of Jamaica, takes notice of a plant which he calls a yellow wall-flower, with a *polygala* leaf; the leaves are like the common milkwort; it hath a yellow tetrapetalous or four-leaved flower, and a small pod. It is much of the nature of the English wall-flowers, which are said to cleanse the liver and reins from obstructions, provoke the menses, and expel the fecundines and dead child.

WALNUTS.

We are not without walnuts in America, especially in Virginia; one fort is called *hickory*. But Sir H. Sloane speaks of two or three forts in Jamaica. I saw
one

one fort growing in Guanaboa, or Golden-Vale, in St. John's parifh, in Jamaica: I obferved its outward fhell was quadrangular, of a yellowifh-green colour, and, when that was taken off, there were four black round kernels, but very white within and pleafant, eating like a filbert; they fay they eat well roafted, as well as raw. I could get nobody to tell what they called them, but one affirmed to me it was *Virginia bread-nut.*

WATER-APPLE.

Some call them Sweet-Apple. I have feen of them very large. Pifo places them among his poifon-plants, but the alligators eat of them, they growing always by river-fides. I have tafted of them, and they feem to have a fweetifh tafte, but are very watery; it may be, the great coldnefs and moifture may make them a fort of poifon to the ftomach.

WATER-CRESSES

Grow in moft fprings and rivers in Jamaica, and the very fame fort as grow in Europe; but, if any thing, thefe in Jamaica are the ftrongeft, and moft peccant and biting upon the tongue.

WATER HEMP-AGRIMONY.

Sir H. Sloane makes two forts of them, and calls them *Eupatorium aquaticum duorum generum,* of which, he faith, we have two forts of our own land, meaning England, and another alfo brought from America; being in all other things very like one unto the other, but only in the placing or fetting of the leaves upon the ftalks, which, in one fort, hath divers leaves fet together, like the figure of a hand, all meeting together at the bottom, fet by diftances at the ftalks, every one

not

not divided but whole, yet dented about the edges, and in form and greenneſs like unto the leaves of wild hemp: And, in the other, which is that we are writing of, every leaf is ſomewhat divided, three or five upon a ſtalk, two at a joint; the flowers are yellow-iſh-brown, made of many leaves like a ſtar, ſet about a middle thrum, with green heads or capſula under them, divers ſtanding together, thruſting forth from the joints with leaves and the tops of the branches, which turn into long flat rugged ſeed, and will ſtick like burs to any garments. The whole plant is ſomewhat aromatic, and taſting ſomewhat ſharp like pepper, and ſo doth the root alſo. Although all theſe ſorts of hemp-like agrimony uſually grow by water-ſides, yet they will grow in drier places. The hemp-like agrimony, or *Eupatorium cannabinum*, is of the ſame temperature of heat and drying as the other ſorts, as opening, cleanſing, and cutting viſcous humours, and therefore good in the jaundice, dropſies, hardneſs of the ſpleen, &c. The juice of it drank is commended againſt inward impoſthumes, and for outward ſwellings applied as a poultice; they provoke urine and the *menſtrua*; a bath of the whole herb is good againſt leproſies, itch, and ſcabs, and is a good vulnerary.

WATER-LILIES.

There are ſeveral ſorts of water-lilies, the roots of which are ſaid to be an antidote againſt the biting of the ſnake called *cobra capella*, or hooded ſnake. The leaves, ſtalks, and flowers of the other water-lilies are good againſt inflammations, hot pains, burnings, or ſcaldings; the oil, anointed on the temples, cauſes reſt; the ſeeds and roots are uſeful in dyſenteries, diarrhœas, gonorrhœas, and weakneſs in women. The
Egyptians

Egyptians make their *fcarbet nufar* of it; the Turks make an infufion of the flowers in water, over-night, to drink the next morning, to keep them from the head-ache. A fyrup of the flowers or conferve is good againft fpitting of blood; and the powder of the feed, given in conferve of hips, does the fame, and is good againft inward heats.

WATER-WITHE.

Some call them wild vine; and indeed this may be called the *true travellers joy*, to thofe that travel the woods, and meet with them, as they will find refrefhment by them; for, by cutting off a piece about a yard long, holding it up, and fucking one end, a great deal of refrefhing water will come into the mouth, and that no fmall quantity, to admiration, as the hunters of wild hogs have often affirmed to me. At one time of the year, it is full of a fort of fmall black grapes, as they call them, but more like currants, and no bigger than elder-berries, growing in bunches almoft like them: I have eat many of them with pleafure.
See Grapes.

WHITE MASTICK.

I met with a great many of thefe trees in falling a piece of ground in the mountains above Guanaboa, in the parifh of St. John. I obferved, they bore a fruit much of the fhape and bignefs of cafhew-ftones, and the gum that came out of it was in fmall little drops, white, and of the fcent of maftick, for which reafon the tree is called fo; and I believe it is as good as any maftick whatever, and of the fame virtues.

WHITE WOOD.

There is a particular tree in Jamaica whofe wood is
fo

fo very white, it is diftinguifhed from other woods by the name of white wood, and is very often called white fiddle-wood.

WILD GINGER

. Grows three or four feet high, with a round ftalk, and covered with long leaves from top to bottom, about four inches long and two broad, graffy and thin, with a great many ribs, like long or rib plantain. The flowers ftand on top of the fpiked ftalk very beautifully, of a pale-purple colour, in which is contained the feed; the root differs much from the other ginger, and is compofed of a great many white, round, thick fibres, about two inches long, fmelling like ginger, and very hot and biting. It purges ftrongly, and is faid to cure cancers.

WINTER CHERRIES.

1. Thefe we have in great plenty in moft parts of America. Sir Hans Sloane, in his Natural Hiftory of Jamaica, tribes them among the nightfhades, having a fcent like them, and having a leaf like the common Englith nightfhade. I never could obferve any difference in the fruit of this and thofe in England.

2 Another fort, which differs from the Englifh only in the colour of its fruit, which is yellow when ripe, as the other is red.

3. The third fort differs from the Englifh, in that the fruit is larger; and, when ripe, is always green; the Englifh always red.

4. There is alfo a leffer fort, with a greenifh fruit.

5. Winter cherries with a white flower, and its bladder or hufks from a red inclining to a greenifh-yellowifh colour, and a yellowifh fruit inclining to red.

The virtues of thefe are nearly one and the fame,

being

being great aperitives and diuretics, the berries being bruifed and fteeped in white wine or rhenifh; the juice, thickened to the confiftence of an extract, has the fame virtues; alfo, four or five berries, bruifed in an ordinary emulfion, wonderfully helps the ftrangury and all ftop-pages of urine. There are troches of winter-cherries, which Lemery hath given an excellent account of their virtues and dofe, which is a drachm: The juice of the leaves and fruit, mixed with Indian pepper, immedi-ately eafes the cholic and provokes urine, and opens all obftructions. There is alfo in South-America a purple-bladder nightfhade; they boil three or four of its ber-ries in white wine or water, and drink it; it is wonder-fully fuccefsful in ftoppages of urine, and in the gravel.

WINTER-GREEN.

There is a plant growing in Brafil called winter-green, with chick-weed flowers; it is cooling, drying, and aftringent, which makes it an excellent wound-herb; it makes an excellent balfam, with hogs fat and turpentine; the juice or the decoction of it is excel-lent for inward wounds or bruifes, and alfo ftops fluxes.

WINTER'S BARK.

This plant grows in great plenty in moft parts of America, and hath the name from one captain Wil-liam Winter, who accompanied Sir Francis Drake in his voyage to America, and, on his return, was the firft that brought it into England, in the year 1579. They found it to be a fingular thing againft the fcurvy, which they were much fubject to on board their fhips. Its leaves are always green and glaffy, like the laurel kinds, but fmaller and rounder, with an aromatic fmell and fpicy tafte; the berries, which are of the big-

O   nefs,

nefs, fhape, and tafte of cubebs, contain a fmall black triangular feed, as hot as the prickly yellow wood feed. The bark of the body of the tree is very thick, and of a dark-whitifh or brown colour without-fide, but whiter within; but I have had fome of the bark pulled off from the fmall branches or limbs, and took care to cure it without any wet or moifture coming upon it, which hath been very white, thin, and much different in tafte from the other bark, not fo hot, but more like the true cinnamon. The powder of it, fnuffed up the roftrils, draws away rheum and moifture, purging the head, and eafing the pain thereof; fprinkled upon old ulcers, it cleanfes and heals them. I look upon it to be more carminative and ftomachic than the true cinnamon, and more proper for the cholic, it being not fo binding.

There is alfo another tree, whofe bark was brought to me by a negro, which was much thinner and redder, coming nearer to the true cinnamon, whilft frefh gathered; but I obferved, as it dried its fcent and tafte feemed to be in a manner loft, and therefore had no further fearch or enquiry after it; but I have confidered fince, that it might be owing to the curing of it.

WITHES.

The number and variety of withes is fo great, that it is in a manner impoffible to give a diftinct account of them. The moft noted for ufe of tying things together, are the prickly-pear withe, the China withe, the pudding withe, &c: befides which there are great numbers of others; one whereof proceeds from a gum-tree. They fall from the boughs, one hanging by another till they touch the ground, from whence they receive fome nourifhment, which makes them grow larger; and if it happen that three or four

of

of them come down fo near one another as to touch, and the wind twift them together, they appear fo like ropes as they cannot be difcerned five paces off whether it be a rope or withe. Thefe are of ufe to the hunters, and thofe who go after rebellious negroes, to help them to climb up the rocks, which in fome places they could not attempt without thefe withes, which come from the trees, which they hold to climb on, and bear any weight.

WOLF'S BANE.

We have a fort of wolf's bane in America; it is a poifon-plant.

WOUND-WORT.

Parkinfon writes of dorias wound-wort, a fort of which grows in America; it heals all wounds and ulcers, inwardly and externally.

XIPHION.

This is a name which Plumier makes ufe of for a plant which he calls *xiphion flore e luteo-nigricante*. I cannot tell what he means, unlefs he means that which is commonly called corn-flag; and if fo, it muft be a fweet-fcented one, and of the kind of *acorus*, five *calamus aromaticus*.

YAMS.

This is one of our principal bread-kinds in Jamaica, of which there are feveral forts, as there are of the potatoes; *viz*. The purple yam: Two forts of white, one of which is called the feed-yam, which is extraordinary white, and makes an admirable fine flour for making of bread or puddings, and thickening broth: Another fort, of a coarfe fulphur-colour or yellowifh yam, called

O 2

negro

negro-yam, whofe ftalks are prickly, and are of the convolvulus kind; the root is a foot or more long, brown on the outfide, and much refembles the common briony-root: One fort of a purplifh colour, and fome of thefe roots are as big as the calf of a man's leg, fome long, fome rounder, and fome flat like a foot, with knobs like toes; the ftalk is of the bignefs of a goofe-quill, fquare at each corner, having a thin reddifh extant membrane, making it alated; it will turn and wind round any thing it comes near, rifing nine or ten feet high, and putting forth leaves at every three inches diftance, fet oppofite to one another, having foot-ftalks two inches long; the leaves are two inches and an half long, and an inch and three quarters broad at the round bafe, almoft in the fhape of an heart and pointed, of a yellowifh-green colour, having many ribs, taking their beginning from the foot-ftalk as from a common centre, with tranfverfe ones between; *ex alis foliorum* come inch-long ftrings, with fmall flowers of a yellowifh-green colour, to which follow many dark-brown feeds of an irregular fhape; but the feed is never planted, but by pieces of the root, which we plant about January or February, and they are fit to dig about Chriftmas. The juice of the leaves is good againft fcorpions fting, and makes good fomentations to cleanfe and heal ulcers.

Yellow Mastick.

It is a hard yellow wood, like box, as durable, and hath alfo the fame fort of leaves.

End of Barham's Manuscripts.

LINNÆAN

LINNÆAN INDEX.

Author's Names	Linnæan Names
ALDER-tree	
Alder-tree, or button-wood	*Conocarpus erecta*
Alligator-wood	*Elutheria*
Alfines, or chick-weed	*Holosteum cordatum*
Ambergris	*Ambra ambrofiaca*
Amber, liquid	
Anchoaca	
Anchovy-pear	*Grias cauliflora*
Angelyn-tree	*Geoffreya inermis*
Anotto	*Bixa orellana*
Apples	
Apples of love	*Solanum lycoperficum*
Apples caufing madnefs	*Solanum melongena*
Apples, thorny	*Datura ftramonium*
Araquidna	*Arachis hypogæa*
Arraganas	
Arrow-head	*Sagittaria lancifolia*
Arrow-root	*Thalia geniculata*
Arfmart	*Polygonum hydropiper*
Afparagus	*Afparagus officinalis*
Attao	*Caffia viminea ?*
Avens	
Avocado-pear	*Laurus Perfea*
Balfams and gums	
Balfam capaiba	*Copaifera officinalis*
Balfam herb	*Dianthera Americana*
Balfam nervinum	
Balfam Peru	*Myraxylon Peruiferum*

Balfam

Author's Names	Linnæan Names
Balsam Tolu	*Tolusera balsamum*
Balsam-tree	*Bursera gummifera*
Banana-tree	*Musa sapientum*
Barbadoes flower fence	*Poinciana pulcherrima*
Basil	*Ocymum basilicum*
Bastard cedar	*Theobroma guazuma*
Bastard mammee, or Santa Maria	*Calophyllum calaba*
Bdellium	
Beans and pease	
Bean-tree	*Erythrina corallodendron*
Belly-ache weed	*Jatropha gossypifolia*
Bignonia	*Bignonia*
Bind-weeds	
Birch-tree	*Bursera gummifera*
Bisnagus, or visnaga	*Daucus visnaga*
Bitter-wood	*Xylopia glabra*
Black mastick	
Blood-flower	*Asclepias Curassavica*
Boxthorn	
Brasilletto	*Cæsalpinia Brasiliensis*
Bread-nut tree	*Brosimum alicastrum*
Brier-rose of America	
Briony	
Brook-lime	
Broom-weed	*Calea scoparia*
Buck-wheat	*Polygonum scandens*
Bully-tree	*Achras salicifolia*
Cacao	*Theobroma cacao*
Calabash	*Crescentia cujete*
Calavances	
Caltroppe	*Tribulus maximus*
Campions	
Canes	*Saccharum officinale*
	Capsicum

Author's Names	Linnæan Names
Capsicum peppers	*Capsicum*
Carapullo	
Cardamon	
Cashew	*Anacardium occidentale*
Cassada	{ *Jatropha manihot* { *Jatropha multifida*
Cassia fistula	{ *Cassia fistula* { *Cassia Javanica*
Cedar	{ *Cedrela odorata* { *Juniperus Bermudiana*
Celandine	*Bocconia frutescens*
Centaury	
Cerasee and cucumis	*Momordica balsamina*
Cherry-tree	*Cordia collococca*
Chili cardinal flower	*Lobelia tupa*
China-root	*Smilax pseudo-China*
Cinnamon	
Citrons	
Clary	*Heliotropium Indicum*
Clove-strife	{ *Oenothera octovalvis* { *Oenothera pumila*
Coca	
Cocoons	*Mimosa scandens*
Colilu or culilu	{ *Amaranthus viridis* { *Amaranthus spinosus*
Contrayerva	*Aristolochia odorata*
Coopers withe	
Copal	*Rhus copallinum*
Corals and corallines	
Cotton	*Gossypium Barbadense*
Cotton-tree	*Bombax ceiba*
Cowhage, or cowitch	*Dolichos pruriens*
Currant-tree	*Ehretia bourreria*
Currato	*Agave vivipara*

Author's Names	Linnæan Names
Custard-apple	*Annona reticulata*
Daisy	
Dandelion	*Tussilago uniflora*
Dildoes	{ *Cactus Peruvianus* { *Cactus repandus*
Dodder	*Cuscuta Americana*
Dogsbane	
Dog-stones	*Orchis*
Dog-wood	*Piscidia erythrina*
Dragon's blood	
Ducks meat, or pond-weed	*Lemna minor*
Dumb-cane	*Arum seguinum*
Dwarf-elder	*Urtica grandifolia*
Dying plants	
Ebony	*Aspalathus ebenus*
Eddos	*Arum esculentum*
Elder	*Piper amalago*
Elemi	*Amyris elemifera*
Elm	*Cordia gerascanthus*
Eryngium, or eringo, or sea-holly	*Eryngium fœtidum*
Female fern	*Polypodium*
Fennel	*Anethum fœniculum*
Ferns	
Fig-Trees	*Ficus Indica*
Fingrigo	*Pisonia aculeata*
Flax-weed	
Flea-banes	*Conyza*
Flore de Paraiso, or flower of Paradise	
Floripondio	*Datura stramonium*
Flower-gentle, or amaranthus	*Amaranthus*
Four o'clock flower	*Mirabilis jalappa*

Fox-

Author's Names	Linnæan Names
Fox-glove, or fox-finger, or finger-wort	
Frutex baccifera, or cloven berries	*Samyda pubefcens*
Fumiterry	
Fuftic	*Morus tinctoria*
Gamboge	*Cambogia gutta*
Garlic-pear	*Crateva gynandra*
Germander, or water-germander	*Stemodia maritima*
Ginger	*Amomum zingiber*
Gland-flax, or nuil	
Golden-rod	*Conyza lobata*
Goofeberry	*Cactus perefkia*
Goofe-foot, or fowbane	*Amaranthus polygonoides*
Goofe-grafs	*Valantia hypocarpia*
Courds	*Cucurbita*
Granadillas	*Paffiflora quadrangularis*
Grapes	{ *Vitis labrufca* { *Coccoloba uvifera*
Graffes	
Green withe	*Cactus aphylla*
Ground-ivy	*Hedera terreftris*
Groundfel	
Guavas	*Pfidium pyriferum*
Guinea-corn, or **panicum**	*Holcus forghum*
Guinea-hen weed	*Petiveria alliacea*
Gum animi	
Gum cancamum	
Gum caranna	
Hare's ears	
Harillo	
Hart's tongues	
Hawk-weed	
Hedge-hyffop	*Helichryfum,*

Author's Names	Linnæan Names
Helchryfum, or golden udweed, golden tufts, or locks	Conyza virgata
Hercules	Zanthoxylum C. Herculis
Hog-gum	Rhus metopium
Hog-weed	Boerhaavia diffusa
Holly-rofe, or fage-rofe	Turnera ulmifolia
Honeyfuckle, or upright woodbind	
Horfe-tail	Equifetum
Hound's tongue	
Indian fhot	Canna Indica
Indigo	{ indigofera tinctoria indigofera argentea
Ipecacuanha	Pfychotria emetica
Iron-wort	Clinopodium vulgare
Jaborand	Piper reticulatum
Jalap	Convolvulus jalapa
Jeffamin	{ Plumieria alba Coffea occidentalis
Ketmia	
Lacayota	
Lagetto tree	Daphne lagetto
Lance-wood	Erythroxylum
Laurels	
Lavender	
Lemons	Paffiflora maliformis
Lentils	
Licti, or luifi plant	
Lignum aloes	
Lignum rhodium, or rofe-wood	Amyris balfamifera
Lignum vitæ	Guaiacum officinale
Lilies	

Author's Names	Linnæan Names
Line, or linden-tree	
Limes	*Citrus medica*, var.
Liquid amber	
Liquorice	{ *Glycine abrus* { *Scoparia dulcis*
Liuto	
Liver-wort	*Lichen*
Locus-tree	{ *Malphigia craffifolia* { *Hymenea courbaril*
Logwood	*Hæmatoxylum Campechianum*
Loofe-ftrife	*Oenothera*
Love-apples	*Solanum lycoperficum*
Lucimo	*Mammea Americana*
Macaw-tree	*Cocos Guineenfis*
Mad apples	*Solanum melongena*
Maguey	*Bromelia karatas*
Mahots	*Hibifcus*
Maiden-hairs	*Adiantum*
Majoe, or macary bitter	*Picramnia antidefma*
Mallows	
Mammee-fapota	*Achras fapota*
Mammee-tree	*Mammea Americana*
Manchioneel	*Hippomane mancinella*
Mangrove-tree	{ *Rhizophora mangle* { *Conocarpus erecta*
Maple	
Marigolds	
Marfh-trefoil, or buckbanes	
Maftick	
Melons	{ *Cucumis melo* { *Cucurbita citrullus*
Milk-wood	*Brofimum fpurium*
Milk-wort	*Polygala paniculata*

Mint

Author's Names	Linnæan Names
Mint	*Ballota suaveolens*
Misletoes	*Viscum verticillatum*
Moon-wort	
Money-wort	
Mosses	
Mouse-ear	
Mug-wort	*Parthenium hysterophorus*
Mulliens	
Mushrooms	{ *Agaricus* *Clathrus cancellatus*
Musk-mallow	*Hibiscus abelmoschus*
Musk-wood	*Elutheria*
Mustard	{ *Cleome spinosa* *Cleome triphylla*
Myrtles	
Nahambu, or nhambi	
Naseberry-tree	*Achras sapota*
Navel-wort	*Hydrocotyle umbellata*
Nephritic-tree	*Mimosa unguis-cati*
Nettles	*Urtica*
Nhandiroba, or ghandiroba	*Fevillea cordifolia*
Nickers	{ *Guilandina bonduc* *Guilandina bonduccella*
Nightshades	*Solanum*
Oak of Cappadocia	
Oil-nuts	*Ricinus communis*
Oily pulse	*Sesamum orientale*
Okra	*Hibiscus esculentus*
Old mens beard	*Tillandsia usneoides*
Oleander, or rose-bay	*Nerium oleander*
Olives	*Bucida buceras*
Onagra	*Mentzelia aspera*
Onobrychis, or cock's head	*Hedysarum*
Opuntia	*Cactus*

Oranges

Author's Names	Linnæan Names
Oranges	*Citrus*
Ortigia	*Loofa hispida*
Osmundas	*Osmunda*
Oyster-green	*Ulva lactuca*
Paica julla	
Pajomirioba	{ *Cassia occidentalis* { *Cassia obtusifolia*
Palghi	
Palqui	
Palms	{ *Phœnix dactylifera* *Elais Guincensis* *Areca oleracea* *Cocos nucifera* *Thrinax parviflora* *Chamærops humilis* *Cocos aculeata*
Panke	
Papaws	{ *Carica papaya* { *Carica posoposa*
Paraguay tea	*Cassine Peragua*
Passion-flowers	*Passiflora normalis*
Payco herba	
Peach-tree	*Amygdalus Persica*
Pease	
Pellitory of the wall	
Penguins	*Bromelia penguin*
Pennyroyal	
Pepper-grass	*Lepidium Virginieum*
Peppers	{ *Piper aduncum* { *Piper umbellatum* { *Piper amplexicaule*
Peumo	
Physic nuts	{ *Jatropha curcas* { *Jatropha multifida*

Piemento

Author's Names	Linnæan Names
Piemento	*Myrtus pimenta*
Pigeon-peafe	{ *Cytifus cajan* *Paullinia Curaffavica* *Paullinia pinnata*
Pilewort	
Pillerilla	*Ricinus communis*
Pilofella	
Pimpernell	*Corchorus filiquofus*
Pindalls	*Arachis hypogæa*
Pine-apple	*Ananas*
Pinks	
Plantain	{ *Sagittaria lancifolia* *Alifma cordifolia*
Plantain-tree	{ *Mufa Paradifiaca* *Heliconia bihai*
Plum-trees	{ *Spondias mombin* *Spondias diffufa* *Spondias myrobalanus* *Chryfobalanus icaco* *Spathelia fimplex*
Poifon berries	*Ceftrum nocturnum*
Polypodium	*Polypodium*
Pomegranates	*Punica granatum*
Pond or river weed	
Popes heads	*Cactus melocactus*
Poponax	*Mimofa juliflora*
Poppy	*Argemone Mexicana*
Poquet	
Potatoes, or batatas	*Solanum batatas*
Prickly white wood	
Prickly withe	*Cactus triangularis*
Prickly wood	
Prickly yellow wood	*Zanthoxylum C. Herculis*
Pumkin	*Cucurbita*

Purflane

Author's Names	Linnæan Names
Purflane	*Portulaca oleracea*
Quamoclit	*Ipomoea quamoclit*
Quefnoa, or quina	
Quillay	
Quinchamali	
Quinquina	*Cinchona officinalis*
Ragwort	
Ramoon	*Trophis Americana*
Rampions	
Raquette	*Cactus Peruvianus*
Reeds	
Reilbon	
Reft-harrow	
Rice	*Oryza fativa*
Ricinus	
Rocket	
Rofemary	*Croton cafcarilla*
Rouncevals	
Rue	
Rupture-wort	*Parietaria microphylla*
Rufhes	{ *Cyperus odoratus* *Cyperus articulatus* *Typha latifolia*
Saffron	*Carthamus tinctorius*
Sage	{ *Lantana annua* *Varronia globofa*
St. John's wort	
Solomon's feal	
Sampier	*Sefuvium portulacaftrum*
Sargaffa, or zargaffo	*Fucus natans*
Sarfaparilla	*Smilax farfaparilla*
Saffafras	*Laurus faffaphras*
Savanna-flower	*Echites umbeliata*
Scabious	*Elephantopus fcaber*
	Scammony

Author's Names	Linnæan Names
Scammony	*Convolvulus Brasiliensis*
Scordium, or water-germander	
Scotch grafs	*Panicum latifolium*
Sea-feather, or fea-fan	*Gorgonia flabellum*
Self-heal, or alheal	*Ruellia paniculata*
Semper vive	*Aloe perfoliata*
Senfible plant	*Mimofa*
Septfoil, or tormentil	
Shaddock	*Citrus decumana*
Silk-grafs	*Bromelia karatas*
Soap-berries	*Sapindus faponaria*
Sorrel	*Ciffus acida* *Hibifcus fabdariffa*
Sour-fop	*Annona muricata*
Spanifh arbour-vine	*Ipomoea tuberofa*
Spider-wort	*Commelina communis* *Commelina zanonia*
Spikenard	*Ballota fuaveolens*
Spirit-leaf	*Ruellia clandeftina*
Spleen-wort	*Afplenium*
Spunges	
Spurges	*Strumpfia maritima* *Euphorbia hypericifolia* *Euphorbia myrtifolia* *Euphorbia maculata*
Star-apple	*Chryfophyllum cainito*
Star-flones	
Star-wort	*Conyza*
Stockvifhhout	*Cæfalpinia veficaria*
Stœchas	*Gnaphalium albicans*
Strawberries	*Fragaria*
Sun-flower	*Helianthus*
Supple-jack	*Paullinia triternata*

Sweet-fop

Author's Names	Linnæan Names
Sweet-fop	Annona fquamofa
Sweet-wood	Laurus
Sycamore	Banifteria laurifolia
Tacamahac	Populus tacamahac
Tamarinds	Tamarindus Indica
Tayo	
Tea	{ Capraria biflora { Thea bohea
Thorny apples	Datura framonium
Thoupa	
Throat-wort	
Toad-flax	
Tobacco	Nicotiana tabacum
Tooth-wort	Plumbago fcandens
Travellers joy	Clematis dioica
Tree-rofemary	
Trumpet-tree	Cecropia peltata
Turnfoles	{ Heliotropium gnaphalodes { Heliotropium Curaffavicum
Valerian	Boerhaavia fcandens
Vanillas, or banillas	Epidendrum vanilla
Vervain	Verbena Jamaicenfis
Velvet-leaf	Ciffampelos paricra
Vines	
Violets	Pfychotria herbacea
Viravida	
Virginia fnake-root	Ariftolochia ferpentaria
Wake robin, or arums	Arum
Wall-flowers	Cleome procumbens
Walnuts	Juglans baccata
Water-apple	Annona paluftris
Water-creffes	Sifymbrium nafturtium
Water hemp-agrimony	Eupatorium
Water-lilies	Nymphæa lotus

P

Water-

LINNÆAN INDEX.

Author's Names	Linnæan Names
Water-withe	Vitis labrusca
White maſlick	
White wood	Bignonia pentaphylla
Wild ginger	Amomum zerumbet
Winter cherries	Phyſalis
Winter-green	
Winter's bark	{ Canella alba
	{ Winterania canella
Withes	Arum funiculaceum
Wolf's bane	
Wound-wort	
Xiphion	Iris martinicenſis
	⌠ Dioſcorea alata
Yams	⎨ Dioſcorea ſativa
	⌡ Dioſcorea bulbifera
Yellow maſlick	

INDEX

I N D E X

DISEASES, REMEDIES, &c.

Dr. BARHAM, *in the foregoing work, mentions, either from his own experience, or the report of others, the following articles*

[No. I.]

As affording remedies for

AGUES---China-root, Peppers, Ragwort, Saffafras, Viravida.

ANEURISMS---Plantain.

ANUS, *discharges of blood from the*---Flower-gentle.

------- *extension of the*---Nightshades.

------- *falling out of the*---Pilewort, Plantain, Vervain.

ASTHMAS. *See* CONSUMPTIONS.

BARRENNESS---Ambergris, Mint, Musk-mallow.

BLADDER. *See* STONE, GRAVEL, *infra;* DIURETIC, No. II.

------------ *ulcerated*---Semper vive.

BLEEDING, *inward or outward*---Blood-flower, Horse-tail, Loose-strife, Quinchamali. *See* STYPTIC, No. II.

BLOODY FLUX. *See* DYSENTERY.

BONES, *pains of the*---Piemento. *See* RHEUMATISM.

BOWELS, *weak*---Balsam capaiba, Coopers withe,

Groundfel,

CONVULSIONS, *nervous*---Mint.

CORNS---Cafhew.

COUGHS---Balfam capaiba, Banana-tree, Ground-ivy, Horfe-tail, Liquorice, Mullens, Oily pulfe, Pellitory of the wall, Pigeon-peafe, Polypodium, Scabious, Stœchas.

CRAB-YAWS---Arrow-head.

CRAMPS---Honeyfuckle, Jalap, Mint, Mifletoes, Oak of Cappadocia, Oil-nuts. *See* JOINTS, *ftiff.*

CUTANEOUS DISEASES---Fumiterry, Muftard, Pepper-grafs.

DEAFNESS---Muftard, Oily pulfe.

DEFLUXIONS---Balfam Tolu, Banana-tree, Box-thorn, Brafilletto, Star-wort, Stœchas.

DIABETES---Indian fhot.

DIARRHŒAS---Ipecacuanha, Water-lilies. *See* EVACUATIONS, *too-liberal.*

DROPSIES---Bean-tree, Belly-ache weed, Capficum peppers, Cafhew, Contrayerva, Dumb-cane, Dwarf-elder, Flax-weed, Manchioneel, Marfh-trefoil, Nettles, Oak of Cappadocia, Oil-nuts, Pellitory of the wall, Pepper-grafs, Peppers, Peumo, Plantain, Ricinus, Rofemary, Scammony, Spurges, Trumpet-tree, Vervain, Water hemp-agrimony.

DRY BELLY-ACHE---Ambergris, Attao, Caffada, Oil-nuts, Spurges *(fpecies 5).*

DYSENTERY---Anotto, Campions, Cotton, Cotton-tree, Flea-banes, Ipecacuanha, Logwood, Loofe-ftrife, Purflane, Tamarinds, Water-lilies.

EAR-ACHE---Garlic pear, Indian fhot.

EMPYEMAS---Oak of Cappadocia.

EVACUATIONS, *too-liberal*--- Ambergris, Blood-flower, Capficum peppers, Palms, Pimpernell, Trumpet-tree.

EXCORIATIONS---Horfe-tail.

EYES.

EYES, *blood-shot*---Violets.

------- *defluxions of the*---Love-apples. *See* DE-
FLUXIONS.

------- *films on the*---Celandine, Papaws, Poppy, Spurges, Vines.

------- *sore*---Balsam-herb, Bean-tree, Boxthorn, Brafilletto, Gourds, Hawk-weed, Loose-strife, Marigolds, Nightshades, Oily pulse, Pigeon-pease, Poppy,
Purslane, Vervain, Vines.

FALLING SICKNESS---Misletoes, Nickers, Star-
wort.

FELONS---Arsmart.

FEVERS---Ambergris, Attao, Brasilletto, Bully-tree,
Centaury, Cerasee, Cherry-tree, Gourds, Grana-
dillas, Lemons, Melons, Nightshades, Oil-nuts,
Penguins, Purslane, Shaddock, Sorrel, Tamarinds.

---------- *hectic*---Anotto, China-root, Okra.

---------- *intermitting*--Centaury, Locus-tree.

---------- *malignant*---Arrow-root, Balsam Peru, Cedar, Contrayerva, Dandelion, Pimpernell, Spikenard, Spurges, Wake robin.

FISTULA IN ANO---Liquid amber.

FISTULAS---Flax-weed.

FITS OF THE MOTHER---Ambergris, Lavender.

FLUXES---Duck's meat, Flea-banes, Flower-gentle,
Germander, Golden-rod, Goose-grass, Grapes,
Guavas, Hawk-weed, Helichrysum, Holly-rose,
Ipecacuanha, Iron-wort, Logwood, Mangrove-tree,
Money-wort, Mulliens, Onobrychis, Palms, Plan-
tain, Plantain-tree, Pond or river weed, Poponax,
Stœchas, Trumpet-tree, Violets, Winter-green. *See*
DYSENTERY.

FRACTURED BONES---Cotton-tree.

FRECKLES. *See* COSMETIC, No. II.

GALL. *See* OBSTRUCTIONS.

GLEETS

GLEETS---Blood-flower.

GOUT---China-root, Cowhage, Currato, Fuftic, Gum caranna, Hog-gum, Marfh-trefoil, Muftard, Oyfter-green, Peppers, Pigeon-peafe, Sarfaparilla, Thorny apples.

--------- knotty---Arfmart.

GRAVEL---Anotto, Arfmart, Capficum Peppers, Currato, Gland-flax, Mallows, Nephritic-tree, Okra, Pellitory of the wall, Spikenard, Spleen-wort, Winter-cherries.

GREEN SICKNESS---Contrayerva.

--------- WOUNDS---Baftard mammee, Goofe-grafs, Harillo, Hog-gum, Self-heal. See WOUNDS.

GUINEA-WORM---Oil-nuts.

HÆMORRHOIDS. See PILES.

HEAD-ACHE, &c.---Ambergris, Attao, Bafil, Garlic pear, Mifletoes, Muftard, Oil-nuts, Onobrychis, Peppers, Purflane, Tacamahac, Thorny apples, Water-lilies, Winter's bark.

HEART-BURN---Pigeon-peafe.

HERNIA CARNOSA. See RUPTURES.

HERPES. See ST. ANTHONY'S FIRE.

HICCOUGHS---Spleen-wort.

HIP---Ambergris.

HOARSENESS---Banana-tree, Canes, Oily pulfe, Palms, Polypodium.

HORSES, galled backs of---Pajomirioba.

HYSTERICS---Buck-wheat, Eryngium, Liquid amber, Rue.

ILIAC PASSION---Peppers.

IMPOSTHUMES---Capficum peppers, Marigolds, Mifletoes, Oak of Cappadocia, Oily pulfe, Water hemp-agrimony.

INFANTS, difeafes of---Liquorice, Oranges, Peach-tree, Penguins.

IN-

MEAGRIM---Nickers.

MELANCHOLY---Ambergris, Polypodium.

MENSES, *immoderate*. See EVACUATIONS, *too-liberal.*

MERCURIAL POISON---Indian fhot.

MESENTERY. *See* OBSTRUCTIONS.

MORBIFIC TAINTS, &c.---Ambergris, Ipecacuanha.

MOUTHS, *diftorted*---Nickers.

---------- *fore*--- Fuftic, Golden-rod, Iron-wort,
Line, Liquid amber, Mulliens, Penguins, Ragwort, Rampions, Self-heal, Vines.

NAILS *of the hands and toes, overgrowing fkins of the*
---Vines.

NERVES, *contracted*---Thorny apples.

---------- *dried*---Oily pulfe.

---------- *weaknefs of the*---Liquid amber, Tacamahac.

---------- *wounded*---Balfam capaiba, Balfam Peru.

NIPPLES, *fiffures or cracks of the*---Nightfhades.

NUMB PALSY---Capficum peppers, Saffafras.

OBSTRUCTIONS---Avens, Balfam capaiba, Cerafee,
Contrayerva, Coopers withe, Dodder, Fumiterry,
Germander, Graffes, Gum cancamum, Maiden hairs,
Navel-wort, Nephritic-tree, Nightfhades, Onobrychis,
Peppers, Polypodium, Reft-harrow, Rue, Saffafras,
Semper vive, Spikenard, Spleen-wort, Tamarinds,
Vanillas, Vervain, Wall-flowers, Winter-cherries.

PALSIES. *See* LIMBS, *cold, weak,* &c.

PESTILENTIAL DISEASES---Germander, Graffes,
Oranges, Pimpernell, Rue, Scabious, Wake robin.

PHLEGM---Canes, Cardamon, Polypodium, Saffron.

PHRENSIES---Nightfhades, Spurges.

PHTHISICS---Balfam Tolu, Pimpernell.

PILES---Blood-flower, Flax-weed, Garlic pear, Mulliens, Nightfhades, Palms, Panke, Pilewort, Toad-
flax, Vervain, Vines.

PLAGUE

PLAGUE---Contrayerva, Dandelion, Pimpernell, Rue, Scabious, Wake robin.

PLEURA, *pains in the*---Germander, Pellitory of the wall.

PLEURISIES---Avens, Centaury, Milk-wort, Misletoes, Oily pulse, Payco herba.

POISONS---Ambergris, Anotto, Arrow-root, Bdellium, Contrayerva, Ginger, Grasses, Jaborand, Lignum aloes, Mustard, Nahambu, Navel-wort, Nhandiroba, Onobrychis, Pajominioba, Peppers, Rue, Rushes, Scordium, Spikenard, Spurges (*species* 5), Velvet-leaf, Wake robin.

POLYPUS---Wake robin.

PURGINGS. *See* EVACUATIONS, *too-liberal.*

QUINSIES---Helichryfum, Liquid amber, Ragwort, Star-wort.

REINS. *See* OBSTRUCTIONS.

RHEUMATISMS---Centaury, Fustic, Nhandiroba, Oil-nuts, Peppers, Piemento, Sarsaparilla, Spikenard.

RICKETS---Osmundas.

RING-WORMS---Celandine, Liver-wort, Pajomirioba, Papaws.

RISING OF THE LIGHTS---Oranges.

RUPTURES---Duck's meat, Mulliens, Rest-harrow, Tooth-wort.

ST. ANTHONY'S FIRE---Cashew, Cerasee, Cowhage, Love-apples, Nightshades, Purslane.

SCAB *or* MANGE IN CHILDREN---Broom-weed.

SCABS, *malignant*---Liver-wort, Water hemp-agrimony.

SCALD-HEADS---Palqui.

SCALDS---Peppers, Purslane, Rushes, Water-lilies.

SCIATICA RHEUMATISMS---Liquid amber, Pepper-grafs.

SCURF---Palqui.

SCURVY

SCURVY---Pepper-grafs, Saffafras, Winter's bark.

SIDES, *ſtiches and pains of the*---Germander, Mifletoes.

SINEWS, *contracted*---Liquid amber, Oily pulfe.

SKIN, *difeafes of the*---Clary, Duck's meat, Oily pulfe, Pond or river weed.

SOLDIERS, *difeafe of, called* DIE BRUEN, *when in camps or garrifons*---Self-heal.

SORES,---Bafil, Golden-rod, Mug-wort, Muftard, Pajomirioba, Rampions, Semper vive, Tobacco, Wake robin.

SPASMS---Ambergris, Oak of Cappadocia, Oil-nuts, Oily pulfe, Quefnea.

SPITTING, *great*---Oranges.

---------- *of blood*---Brier-rofe of America, Loofe-ftrife, Plantain-tree, Purflane, Water-lilies.

SPLEEN, *diſtempers of the*---Spleen-wort. *See* OBSTRUCTIONS.

---------- *fwelling and hardnefs of the*---Honeyfuckle, Indian fhot, Maple, Pellitory of the wall, Vervain, Water hemp-agrimony.

STINGS OF SNAKES, SPIDERS, &c.---Arrow-root, Bafil, Clary, Contrayerva, Eryngium, Goofe-grafs, Hare's ears, Ipecacuanha, Nahambu, Pimpernell, Pindalls, Rue, Scabious, Spider wort, Spurges, Turnfoles, Water-lilies, Yams.

STOMACH, *cold, weak, &c.*---Anotto, Balfam capaiba, Balfam Peru, Bitter-wood, Brafilletto, Capficum peppers, Centaury, Contrayerva, Flea-banes, Germander, Ginger, Groundfel, Gum cancamum, Mallows, Marigolds, Muftard, Myrtles, Oily pulfe, Okra, Oranges, Peppers, Piemento, Rufhes, Sage, Semper vive, Vervain.

STONE---Arfmart, Capficum peppers, Currato, Golden-rod, Mallows, Nephritic-tree, Okra, Paraguay tea, Payco herba, Spikenard.

STRAN-

STRANGURY---Anotto, Germander, Goose-foot, Melons, Nightshades, Pellitory of the wall, Purslane, Rice, Spleen-wort, Winter-cherries.

SURFEITS---Semper vive.

SWEATING, *immoderate*---Anotto.

SWELLINGS, *cold, &c.*---Arsmart, Cassada, Duck's meat, Goose-foot, Mallows, Misletoes, Peppers, Plum-trees, Ragwort, Self-heal, Water hemp-agri-mony.

TENDONS, *contracted*---Thorny apples.

TETTERS---Celandine, Liver-wort.

THROATS, *sore*---Fustic, Iron-wort, Liquid amber, Loose-strife, Ragwort, Rampions, Self-heal, Throat-wort, Vervain.

THRUSH---Penguins.

TOOTH-ACHE, *&c.*---Arsmart, Attao, China-root, Coca, Ebony, Guinea-hen weed, Purslane, Tobacco.

TUMOURS---Clove-strife, Ground-ivy, Oily pulse.

----------- *cancerous*---Nightshades.

----------- *cold*---Balsam Peru.

----------- *schrofulous and schirrous*---Arsmart, Mis-letoes.

TYMPANY---Nettles.

ULCERS---Basil, Boxthorn, Cashew, Clary, Dog-wood, Fox-glove, Golden rod, Helichrysum, Her-cules, Hog-gum, Honeysuckle, Horse-tail, Liver-wort, Maiden hairs, Majoe, Mangrove-tree, Myrtles, Oak of Cappadocia, Osmundas, Pajomirioba, Pen-guins, Physic-nuts, Piemento, Sage, Self-heal, Spikenard, Spirit-leaf, Star-wort, Tamarinds, Thoupa, Tobacco, Trumpet-tree, Wake robin, Winter's bark, Wound-wort, Yams.

URINE, *heat of*---Banana-tree, Purslane, Tamarinds.

-------- *stoppage of.* See DIURETIC, No. II.

-------- *viscid or purulent*---Nettles, Sargassa.

VA.

REMEDIES, &c.

VAPOURS---Tacamahac.

VENEREAL CASES---Balfam capaiba, Birch-tree, Blood-flower, China-root, Coopers withe, Elder, Fingrigo, Hog-gum, Lignum vitæ, Limes, Liver-wort, Loofe-ftrife, Majoe, Mallows, Nickers, Oil-nuts, Prickly white wood, Purflane, Sarfaparilla, Saffafras, Spleen-wort, Spurges, Star-wort, Tamarinds, Trumpet-tree, Water-lilies.

VISCERA, *obftructions of the*---Dandelion.

VOMITING. *See* EVACUATIONS, *too-liberal*

WARTS---Celandine, Papaws, Spurges.

WATERY HUMOURS---Caffada, Cerafee, Gamboge, Onobrychis, Peach-tree, Rofemary, Saffron, Spanifh arbour vine, Travellers joy.

WEAKNESS---Ambergris, Dog-ftones.

--------------- *female*---Balfam capaiba, Blood-flower, Liquid amber, Liver-wort, Water-lilies.

WHITLOWS---Arfmart.

WOMB, *hardnefs of the*---Liquid amber, Oily pulfe.

--------- *pains, &c. in*---Oily pulfe, Oranges, Pellitory of the wall, Tacamahac.

WORMS---Angelyn-tree, Bitter-wood, Cafhew, Cedar, Centaury, Female fern, Germander, Graffes, Gum cancamum, Lignum aloes, Locus-tree, Onobrychis, Oranges, Oyfter-green, Penguins, Phyfic-nuts, Rocket, Semper vive, Vervain.

----------- *in cattle*---Semper vive.

WOUNDS---Arrow-head, Avens, Balfam capaiba, Clary, Fox-glove, Golden-rod, Hare's ears, Iron-wort, Liquid amber, Loofe-ftrife, Money-wort, Moufe-ear, Mulliens, Ofmundas, Pigeon-peafe, Pimpernell, Ragwort, Sage, Self-heal, Tooth-wort, Trumpet-tree, Velvet-leaf, Winter-green, Wound-wort.

YAWS

YAWS---Lignum vitæ, Majoe, Oil-nuts. *See* CRAB-Yaws.

YELLOW JAUNDICE---Cerafee, Marigolds, Spleen-wort.

[No. II.]

The following qualities are afcribed to the annexed articles:

ALOETIC---Currato, Silk-grafs.

ANODYNE---Mallows, Nhandiroba, Nightfhades, Peppers, Piemento, Scabious, Tacamahac.

APERITIVE---Apples, Avens, Bdellium, Caffia fif-tula, Centaury, Cerafee, Contrayerva, Dodder, Four o'clock flower, Gamboge, Goofeberry, Goofe-foot, Graffes, Groundfel, Jalap, Lignum vitæ, Locus-tree, Nightfhades *(fpecies* 6), Onobrychis, Ricinus, Saffron, Scabious, Sorrel, Spurges, Tamarinds, Turnfoles, Vanillas, Violets, Water hemp-agrimony, Winter-cherries.

---------------- *and afterwards aftringent and ftrength-ening*---Apples, Ipecacuanha.

ASTRINGENT---Alder-tree, Alder-tree or button-wood, Boxthorn, Brier-rofe of America, Caltroppe, Campions, Dog-wood, Female fern, Ferns, Flea-banes, Flower-gentle, Fuftic, Garlic pear, Golden-rod, Grapes, Guavas, Hawk-weed, Helichryfum, Holy-rofe, Mangrove-tree, Myrtles, Nightfhades, Old mens beard, Olives, Oyfter-green, Palms, Plantain, Pomegranates, Poponax, Rice, Rupture-wort, Rufhes, Septfoil, Stœchas, Tamarinds, Tooth-wort, Trumpet-tree, Winter-green.

ATTENUATING---Avens, Canes, Peppers, Scabious, Vanillas.

BALSAMIC---Loofe-ftrife, Muftard, Peppers.

CAR

CARDIAC---Ambergris, Banana-tree, Contrayerva, Lignum aloes, Oranges, Peppers, Rue, Spikenard.

CEPHALIC---Gum caranna, Lignum aloes, Vanillas, Vervain. *See* HEAD-ACHE, *&c.* No. I.

CLEANSING---Avens, Clary, Ferns, Hercules, Indian fhot, Pajomirioba, Peppers, Pigeon-peafe, Rampions, Scabious, Semper vive, Tobacco, Water hemp-agrimony, Winter's bark.

COOLING---Alder-tree, Alfines, Caltroppe, Duck's meat, Fuftic, Garlic pear, Goofeberry, Hawk-weed, Hog-weed, Indian fhot, Love-apples, Melons, Night-fhades, Okra, Oyfter-green, Pajomirioba, Plantain, Pond or river weed, Popes heads, Purflane, Rampions, Rice, Shaddock, Tamarinds, Tooth-wort, Turnfoles, Valerian, Water-lilies, Winter-green.

COSMETIC---Cacao, Cafhew, Cerafee, Cotton, Cotton-tree, Honeyfuckle, Jeffamin, Loofe-ftrife, Oak of Cappadocia, Purflane, Tamarinds, Travellers joy, Violets, Wake robin.

COUNTER-POISON, *a potent*---Spurges, *fpecies* 5. *See* POISONS, No. I.

DIGESTIVE---Bdellium, Scabious.

DISCUSSIVE---Bdellium, Clove-ftrife, Floripondio, Oily pulfe, Water hemp-agrimony.

DIURETIC---Anotto, Afparagus, Balfam capaiba, Bean-tree, Capficum peppers, Cafhew, Contrayerva, Cowhage, Currato, Eryngium, Flax-weed, Gland-flax, Golden-rod, Graffes, Mallows, Melons, Milkwort, Nephritic-tree, Nightfhades, Okra, Oranges, Penguins, Pepper-grafs, Peppers, Reft-harrow, Rocket, Sampier, Sargaffa, Scordium, Spikenard, Spurges (*fpecies* 5), Vanillas, Vines, Wake robin, Water hemp-agrimony, Winter cherries.

DRYING---Alder-tree, Alder-tree or button-wood, Avens, Brier-rofe of America, **Buck-wheat**, Cam-

phorc,

pions, Celandine, Ferns, Hawk-weed, Helichryfum, Holly-rofe, Moufe-ear, Old mens beard, Oyfter-green, Pond or river weed, Scabious, Scordium, Tooth-wort, Violets, Winter-green.

EMETIC---Belly-ache weed, Cocoons, Gamboge, Goofe-foot, Groundfel, Navel-wort, Ortigia, Phyfic-nuts.

EMOLLIENT---Mallows, Oily pulfe, Okra, Peppers.

FEVERISH---Pumkin, *if eaten too much.*

HEATING---Avens, Buck-wheat, Cacao, Celandine, Cinnamon, Navel-wort, Oily pulfe, Onobrychis, Oranges, Peppers, Prickly white wood, Rofemary, Scabious, Violets.

INCARNATIVE---Semper vive.

INTOXICATING---Carapullo.

MOISTENING---Duck's meat, Hog-weed, Oily pulfe, Purflane, Valerian.

NARCOTIC---Lignum aloes, Pigeon-peafe, Poppy, Rufhes, Thorny apples.

NUTRITIVE---Avocado-pear, Cacao, Calavances, Caffada, Colilu, Eddos *(fome forts)*, Guinea-corn, Mad-apples, Nightfhades *(fpecies 3)*, Okra, Palms, Pigeon-peafe, Pindalls, Plantain-tree, Potatoes, Yams.

PECTORAL---Balfam Peru, Balfam Tolu, Cardamon, Cotton, Cotton-tree, Nettles, Nightfhades, Oily pulfe, Okra, Saffron, Scordium.

POISONOUS---Caffada *(with the antidote)*, Chili cardinal flower, Goofe-foot, Licti *(with its anti-dote)*, Manchioneel, Mufhrooms *(with the antidote)*, Paica julla, Poppy, Savanna-flower *(with the anti-dote)*, Water-apple, Wolf's bane.

PURGATIVE---Belly-ache weed, Caffada, Cocoons, Nightfhades, Oil-nuts, Ortigia, Paica julla, Phyfic-nuts, Quamoclit, Rocket, Spanifh arbour-vine, Spurges, Wild ginger.

SCOR-

SCORBUTIC---(Sugar, if too much ufed, *under the* *article)* Canes.

STOMACHIC---Contrayerva, Coopers withe, Eryn-gium, Ginger, Lignum aloes, Muftard, Onobrychis, Oranges, Scordium, Vanillas, Wake robin, Win-ter's bark. *See* STOMACH, *cold, weak, &c.* No. I.

STYPTIC---Blood-flower, Mangrove-tree, Olives.

SUDORIFIC---Balfam capaiba, Bdellium, Centau-ry, Contrayerva, Flea-banes, Ginger, Payco herba, Sarfaparilla, Spurges, Violets.

VENOMOUS, *if taken inwardly*---(Horfe-beans *and* cocoons, *under the article)* Beans and peafe, Dumb-cane, Oleander.

--------------- *to the eyes*---Chili cardinal flower, Fig-trees, Manchioneel *(third fort)*, Spurges *(fpecies* 1).

VISCOUS---Fingrigo, Milk-wood.

VULNERARY---Alder-tree or button-wood, Balfam capaiba, Cerafee, Daify, Flea-banes, Fox-glove, Gum caranna, Money-wort, Muftard, Pigeon-peafe, Scor-dium, Semper vive, Spirit-leaf, Tobacco, Tooth-wort, Vervain, Water hemp-agrimony, Winter's bark. *See alfo* SORES, ULCERS, WOUNDS, *in* No. I.

[No. III.]

Thefe are reprefented as being of ufe to

ABORTIONS, *prevent*---Plantain.

AFTER-PAINS, *eafe. See* BIRTHS, *&c.*

BIRTHS, *&c. haften, clear, &c.*---Arrow-root, Bdel-lium, Calabafh, Capficum peppers, Flax-weed, Germander, Honeyfuckle, Marigolds, Mint, Oak of Cappadocia, Peppers, Sarfaparilla, Wake robin, Wall-flowers.

BLOOD, *fweeten the*---Balfam capaiba, Centaury Contrayerva, Dandelion, (Docadilla, *under the ar-ticle)* Dying plants, Sarfaparilla.

BONES,

VISCOSITIES *and* TARTAROUS HUMOURS, *diffolve*---Capficum peppers, Goiden-rod, Reſt-harrow.

WIND, *expel*---Bean-tree, Cardamon, Eryngium, Gland-flax, Locus-tree, Mint, Muſk-mallow, Muſtard, Myrtles, Nahambu, Nightſhades, Oranges, Polypodium, Ruſhes, Spikenard, Vanillas, Winter's bark.

WOMENS MILK, *dry up*---Pillerilla, Plantain.

―――――――――――――――― *excite*――― Gland-flax, Pillerilla, Rampions.

[No. IV.]

The following are ſaid to anſwer as ſubſtitutes for

ASPARAGUS, *garden*---Aſparagus.
BROOK-LIME, *Engliſh*---Brook-lime.
CALAMUS AROMATICUS---Ruſhes.
CAMPIONS---Tooth-wort.
CAT-MINT, *Engliſh*---Mint.
CHINA-ROOT, *Eaſt-India*---China-root.
FERNS, *common*---Oſmundas, Polypodium.
FLAX-WEED, *common*---Toad-flax.
GUM ARABIC---Cedar.
―――― GUAIACUM---Manchioneel.
HEMP, *European*---Mallows, Sorrel, Trumpet-tree.
JALAP---Four o'clock flower.
JESUITS BARK---Buſly-tree, Centaury, Locus-tree.
LAND PLANTAIN---Plantain.
LILIES, *European*---Lilies.
LINSEED OIL---Oily pulſe.
LOOSE-STRIFES, *Engliſh*---Looſe-ſtrife.
LUNG-WORT---Tooth-wort.
MARSH-MALLOWS---Okra.
MISLETOES, *Engliſh*---Miſletoes.

Q 2   MONEY-

MONEY-WORT, *English* --Money-wort.

MOSSES, *European*---Mosses.

NETTLES, *English*---Nettles.

OIL OF ALMONDS---Pindalls.

PELLITORY, *European*---Pellitory of the wall.

PERUVIAN QUILL BARK---Locus-tree.

PURSLANES---Alsines, Hog-weed.

RED CORAL---Corals and corrallines.

REEDS, *English*---Reeds.

RHODIUM---Elm.

ROSEMARY, *English*---Rosemary.

SAGE, *English garden*---Sage.

SAMPIER, *English*---Sampier.

SCABIOUS, *Spanish*---Scabious.

SENNA, *Alexandrian*---Barbadoes flower fence.

SUN-FLOWERS, *English*---Sun-flowers.

TEA, *East-Indian*---Tea.

TOBACCO---Coca.

WALL-FLOWERS, *English*---Wall-flowers.

WALNUT-TREE LEAVES, *English*---Cashew.

WILD MARIGOLDS, *European*---Marigolds.

[No. V.]

These are known or supposed proper for

ALOES, *making*---Semper vive.

ARBOURS---Bignonia, Cerasee, Lacayota, Lemons,
Spanish arbour-vine.

ARROWS, *heading*---Palms.

BALSAMS, *making*---Balsam-herb, Balsam nervinum,
Balsam Peru, Spirit-leaf, Tobacco, Winter-green.

BATHS *and* FOMENTATIONS --- Broom-weed,
Coopers withe, Mug-wort, Myrtles, Peppers, Pie-
mento, Plum-trees, Rosemary, Sage, Spikenard,
Spurges, Star-wort, Water hemp-agrimony, Yams.

BED-

BEDSTEADS and PRESSES, *making*---Bitter-wood.

BITTER WINE, *making*---Contrayerva.

BLACK INK, *making*---Poponax.

BLUE, *making*---Indigo.

BOWS, *making*---Macaw-tree.

BROOMS, *making*---Broom-weed.

CABINET WORK---Elm.

CANOES, *making*---Cotton-tree.

CAULKING STUFF, *making*---Palms.

CERGILIM OIL, *making*---Oily pulfe.

CHINKS or BUGS, *keeping away*---Bitter-wood, Sweet fop.

CHOCOLATE, *making*---Cacao, Cafhew, Oily pulfe.

---------------- *ufing in*---Anotto, Vanillas.

CLOTH, *making*---Lagetto-tree, Palms.

COCHINEAL TO FEED UPON---Opuntia.

COCKROCHES, *driving away*---Bitter-wood, Manchioneel.

CORDAGE, *making*---Palms.

CORDIALS---Ambergris, Arrow-root, Bafil, Marigolds, Mufk-mallow, Vanillas.

CUPS, DISHES, SPOONS, &c. *making*---Calabafh, Fig-trees, Palms.

DEAD BODIES, *preferving*---Semper vive.

DISTILLERS USE---Tree-rofemary.

DRINK, *making*---Banana-tree, Cafhew, Potatoes, Sorrel.

DYERS USE---Anotto, Barbadoes flower fence, Brafilletto, Dying plants, Indigo, Logwood, Moffes, Opuntia, Panke, Poponax, Poquett, Reilbon, Saffron, Sorrel, Stockvifhhout.

ELATERIUM, *making*---Cerafee.

FENCES---Barbadoes flower fence, Bean-tree, Limes, Logwood, Nightfhades (*fpecies* 6).

FISHING NETS, *making*---Silk-grafs.

FODDER

FODDER---Baſtard cedar, Bread-nut tree, Guinea-corn, Ramoon, Trumpet-tree.

GARGLES, *making*---Puſtic.

GREEN WALKS---Olives.

GUM, *making*---Fig-trees.

HAMMOCKS, *making*---Silk-graſs.

HATS, *making*---Cotton-tree.

HOOPS, *making*---Coopers withe, Elm.

HOUSES, *caſing*---Palms (*ſpecies* 3).

---------- *covering*---Palms (*ſpecies* 5 *and* 6).

ISSUES, *making*---Caſhew.

JELLIES---Cherry-trees, Sorrel.

LACE, *making*---Lagetto-tree.

LANCES---Lance-wood.

LAPIS CONTRAYERVA, *mixing in*---Arrow-root.

LIME, *making*---Corals and corallines.

LINES, *making*---Currato, Mallows.

LIXIVIUM, *making*---Trumpet-tree.

MANDARIN BROTH, *making*---Oily pulſe.

MEAL, *making*---Caſſada, Liuto, Yams.

MICE, *killing*---Ruſhes.

NICKLACES, *making*---Liquorice.

ODORIFEROUS OIL, *making*---Jeſſamin, Oily pulſe.

OIL, *making*---Phyſic-nuts, Pindalls.

PASTE FOR CONFECTIONARY---Liuto.

PERFUME, *making*---Jeſſamin.

PICKLES---Anchovy pear, Canes, Capſicum peppers, Sampier.

PRESERVES---Arrow-root, Ginger, Gourds, Lacayota, Oranges.

PURGING SYRUP, *making a*---Barbadoes flower fence, Lignum vitæ.

RED INK, *making*---Braſilletto.

ROPES, *making*---Currato, Lagetto-tree, Mahots, Mallows, Silk-graſs, Trumpet-tree.

SAUCES,

SAUCES, *making*---Anotto, Capsicum peppers, Love-apples, Papaws, Sorrel.

SCAMMONY---Spanish arbour-vine.

SILK, *making*---Penguins, Silk-grass.

SOAP, *using as*---Currato, Quillay *(for woollen)*, Soap-berries.

SOUPS *and* BROTHS, *using in*---Anotto, Hedge-hyssop, Okra, Saffron.

SPIRIT, *distilling a*---Calabash, Canes, Cashew, Marsh trefoil, Palms *(species* 4*)*, Rice.

SPOKES FOR WHEELS, *making*---Brasilletto.

STUFFS, *making*---Silk-grass.

SUGAR, *making*---Canes, Palms *(species* 4*)*.

SWEETMEATS, *making*---Palms *(species* 4*)*, Papaws.

SYRUP, *making a*---Marsh-trefoil, Sorrel.

TACKLE FOR SHIPS, *making*---Palms.

TANNING-LEATHER --- Mangrove-tree, Olives, Panke.

TARTS, PUDDINGS, &c. *making*---Banana-tree, Guavas, Papaws, Sorrel.

THREAD, *making*---Maguey.

UNGUENTUM DIALTHEÆ, *making*---Mallows.

VINEGAR, *making*---Penguins.

WINE, *making*---Penguins.

WOOD, *dying of*---Mosses.

[No. VI.]

Under the subjoined heads, mention is made of the trees or plants which produce

BALSAMS---Balsam capaiba, Balsam Peru, Balsam Tolu, Balsam-tree, Basil, Bastard mammee, Bdellium, Fig-trees, Harillo, Liquid amber.

CABBAGE---Palms *(species* 3*)*.

CINNAMON---Winter's bark.

DRINK

DRINK---Grapes, Palms *(species 4)*, Water-withe.

GUM EUPHORBIUM---Raquette.

GUMS---Balfam-tree, Bdellium, Birch-tree, Brafilletto, Cafhew, Cedar, China-root, Copal, Dragon's blood, Elemi, Gamboge, Gum cancamum, Gum caranna, Hog-gum, Lignum vitæ, Mammeetree, Manchionel, Palms *(species 10)*, Tacamahac, White matuck.

JESUITS BARK *or* PERUVIAN BARK---Quinquina.

OIL---Palms *(species 2)*.

SCAMMONY---Scammony.

WAX---Myrtles.

WINE---Palms *(species 2)*.

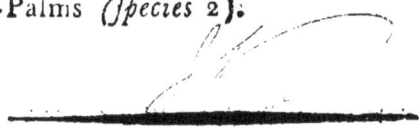

www.ingramcontent.com/pod-product-compliance
Lightning Source LLC
Chambersburg PA
CBHW020848270326
41928CB00006B/606